MOHAMMAD UMAR

THE EARTH'S BEGINNING

A Journey Through Four Billion Years of Fire, Water, and Wonder

First edition

This book was professionally typeset on Reedsy.
Find out more at reedsy.com

Contents

Prologue

A World Before Worlds

Close your eyes and try to imagine nothing.

Not darkness darkness is something, a void you could theoretically fill. Not emptiness emptiness implies space waiting to be occupied. Not silence silence requires the absence of sound, which means sound could exist. Try to imagine the complete absence of everything: no space, no time, no matter, no energy, no possibility of anything at all.

You can't do it. Your mind rebels against the concept because your thoughts themselves require time to unfold, require the electrochemical processes of neurons firing in sequence, require the existence of matter organized into the impossibly complex structure of a human brain. You are made of atoms that obey physical laws in a universe that has structure and history, and from inside that universe, true nothingness is not just unimaginable it's literally inconceivable.

And yet, roughly 13.8 billion years ago, something very much like nothing became something. Space and time themselves began. Energy flooded into existence. The universe was born.

This book is not about that beginning, though we'll start there because we must—the story of Earth begins long before Earth existed, begins with processes operating on scales so vast they make our planet look like a grain of sand on an infinite beach. But what we're really after is more specific, more intimate, more connected to the ground beneath your feet and the air you're breathing right now. We're trying to understand how one particular world, orbiting one ordinary star in one unremarkable galaxy among hundreds of billions, came into being.

We're trying to understand how Earth began.

This is a detective story, but one where almost all the evidence has been destroyed. The first several hundred million years of Earth's history left virtually no direct traces the rocks that formed then have been melted and recycled, subducted back into the planet's interior, metamorphosed beyond recognition by billions of years of heat and pressure. What we know about Earth's beginning comes from fragments: tiny mineral grains older than any intact rock, chemical signatures preserved in younger materials, comparisons with other planets and moons in our solar system, observations of young stars surrounded by disks of gas and dust similar to the one from which our solar system formed, computer simulations of planetary formation, experiments recreating the extreme conditions of the early Earth in laboratories, and the mathematical laws of physics and chemistry applied to conditions that no human has ever witnessed.

From these fragments, we've reconstructed a story. It's a story still incomplete in many details, still debated among scientists, still being revised as new evidence emerges. But the broad outlines are clear, verified through multiple independent lines of evidence, consistent with everything we've learned about how planets form and how matter behaves under extreme conditions.

It's a violent story. Earth was not created peacefully or purposefully but assembled through countless catastrophic collisions, built from the wreckage of smaller worlds that smashed together with enough force to vaporize rock, heated until its entire surface glowed like a star, bombarded for hundreds of millions of years by asteroids and comets, nearly destroyed by an impact that created the Moon. The early Earth was hell by any reasonable definition a molten ball of rock with a toxic atmosphere, subjected to radiation that would sterilize any life that tried to exist, pounded by impacts that released more energy than all nuclear weapons ever built.

Yet from that violence emerged the most beautiful and complex planet we know of in the universe a world with oceans and continents, with an atmosphere that sustains life, with climate systems that have remained stable enough for billions of years to allow life not just to survive but to flourish and diversify into millions of species. The planet that began in fire and chaos somehow transformed itself into a home.

This transformation was not quick. The story we'll tell unfolds over more than a billion years, a span of time longer than the entire fossil record of complex animals, longer than it took for single-celled organisms to evolve into humans. If you compressed Earth's entire history into a single year, the period we'll cover from the formation of the solar system to the establishment of the first stable continents and the origin of life would occupy roughly the first two and a half months. The remaining nine and a half months, containing everything from the evolution of photosynthesis to the rise of dinosaurs to the emergence of human civilization, lie beyond our scope, worthy of their own books.

We'll travel through deep time, visiting epochs when no human eyes existed to witness the events we describe, when the very idea of witnessing anything was meaningless because consciousness itself didn't exist. We'll watch a cloud of gas collapse to form the Sun, see dust grains collide and stick together to build planetesimals, witness the growing Earth being struck by an object the size of Mars in the collision that created the Moon. We'll stand on a surface too hot to solidify, under an atmosphere of rock vapor, watching the first oceans form from millions of years of rain. We'll trace the chemical evolution of increasingly complex molecules in tidal pools and hydrothermal vents, approaching but not quite reaching the threshold of life. And we'll watch as that threshold is finally crossed, as chemistry becomes biology, as Earth transitions from being merely a planet to being a living world.

This is a story told with as much attention to scientific accuracy as current knowledge permits, but also with an eye toward making these vast scales and extreme conditions comprehensible, helping you feel what it might have been like to witness these events even though witnessing them would have been instantly fatal. The numbers involved billions of years, thousands of degrees, trillions of tons of rock quickly become meaningless abstractions unless we can connect them to human experience somehow, find ways to make them visceral and real despite their aliveness.

The story of Earth's beginning is also, in a sense, your story. Every atom in your body was present in that collapsing cloud four and a half billion years ago. The iron in your blood was forged in a supernova explosion before the solar system existed. The calcium in your bones, the carbon in your DNA, the oxygen you breathe all of it has a history stretching back through Earth's formation to the deaths of ancient stars. You are not separate from this story but woven into it, a temporary arrangement of atoms that have been cycling through rocks and water and air and living things for billions of years, briefly taking the form of a consciousness capable of contemplating its own origins before those atoms disperse again to become parts of other things.

Understanding where we came from, how this planet formed and evolved, changes how we see ourselves and our world. It reveals that the blue marble photographed from space is not a static, eternal backdrop for human activity but a dynamic, evolving system that took billions of years to reach its current state and that continues changing even now. It shows us that the conditions allowing our existence liquid water, a protective magnetic field, an atmosphere with oxygen, stable continents, a climate that varies but remains within habitable bounds are not guaranteed features of rocky planets but hard-won outcomes of specific sequences of events, some of them apparently quite lucky.

The beginning of Earth is worth understanding not just as an intellectual exercise, not just to satisfy curiosity about the past, but because it illuminates the present and potentially the future. The processes that shaped our planet accretion, differentiation, out gassing, weathering, the carbon cycle, the origin and evolution of life are still operating, still influencing Earth's trajectory. We live on a planet that is still being formed, still evolving, and for the first time in its history, one species has become capable of influencing those processes on scales comparable to geological forces.

This book is an invitation to journey backward in time, to witness the birth of the world, to understand the deep history that made us possible. It's a journey that requires imagination because the landscapes we'll visit no longer exist, having been transformed by billions of years of change. But it's also a journey grounded in evidence, in the hard-won knowledge accumulated by generations of scientists who have pieced together Earth's story from the fragments that remain.

The story begins not with Earth but with the ending of something else with stars dying so that planets might one day form from their remains.

Introduction

Reading the Unwritten Past

The answer lies in fragments, in clues scattered across the solar system and even beyond, in the laws of physics and chemistry that govern how matter behaves under extreme conditions, and in the ingenuity of scientists who have learned to read stories written in languages most people never knew existed.

Some of the evidence comes from tiny mineral grains called zircons, crystals smaller than sand grains that can survive when the rocks containing them are completely destroyed. Zircons incorporate uranium into their crystal structure when they form, and uranium decays radioactively into lead at a known rate, acting as a geological clock that keeps ticking for billions of years. By measuring the ratio of uranium to lead in a zircon crystal, scientists can determine when that crystal formed with remarkable precision. The oldest zircons ever found, discovered in the Jack Hills of Western Australia within much younger rocks, date back to 4.4 billion years ago just 150 million years after Earth formed, a mere three percent of our planet's total age.

These ancient zircons tell us more than just their age. The oxygen isotopes within them different versions of oxygen atoms with varying numbers of neutrons preserve information about the temperature and chemical environment in which the crystals formed. The ratios suggest these zircons crystallized from magma that had interacted with liquid water, providing some of our strongest evidence that oceans or at least substantial bodies of water existed on Earth as early as 4.4 billion years ago, far earlier than most scientists expected based on models of how long it should have taken for the planet to cool after its violent formation.

Other evidence comes from meteorites fragments of asteroids and sometimes other planets that have fallen to Earth, bringing samples from the early solar system that never experienced the extreme heating and recycling that destroyed Earth's oldest rocks. Some meteorites, called chondrites, contain tiny spherical grains called chondrules that formed when molten droplets of rock solidified in the solar nebula, the disk of gas and dust from which the planets formed. These chondrules date to 4.567 billion years ago, marking the approximate time when solid materials first began condensing in the solar system.

The meteorites also contain presolar grains mineral crystals that are literally older than the solar system itself, formed in the atmospheres of other stars and somehow survived incorporation into the cloud that would become our solar system. We can identify these grains because their isotopic compositions are wildly different from anything that formed in our solar system, bearing the chemical signatures of the nuclear reactions that occurred in the stars where they originated. These microscopic visitors from other stellar systems remind us that Earth's atoms have deep histories extending back through multiple stellar generations.

The Moon provides another crucial source of evidence. Unlike Earth, the Moon is geologically dead it has no plate tectonics to recycle its crust, no atmosphere or liquid water to erode rocks, minimal volcanic activity for the past three billion years. Its surface preserves a record of conditions in the early solar system that Earth's active geology has erased. The lunar highlands, the bright regions visible from Earth without a telescope, consist of ancient rock dating back 4.4 to 4.5 billion years, and the heavily cratered surface records the intense bombardment that both Earth and Moon experienced during the solar system's first several hundred million years.

The rocks brought back by Apollo astronauts between 1969 and 1972 842 pounds of lunar material carefully curated and still being studied decades later have told us about the Moon's composition, its age, and how it formed. The isotopic similarities between lunar rocks and Earth rocks, far closer than between Earth and any meteorite type, strongly support the giant impact hypothesis the idea that the Moon formed from debris thrown into orbit when a Mars-sized object collided with the early Earth. The Apollo samples also preserve a record of the late heavy bombardment, a period around 3.9 to 3.8 billion years ago when the impact rate seems to have spiked dramatically, possibly due to a reorganization of the outer planets' orbits that sent icy bodies from the outer solar system streaming inward.

Observations of other planetary systems provide yet another source of insight. Since the 1990s, astronomers have discovered thousands of planets orbiting other stars, ranging from massive gas giants orbiting closer to their stars than Mercury is to our Sun, to rocky worlds similar to Earth, to systems completely unlike our solar system with architectures we never anticipated. By studying how common different types of planetary systems are, what kinds of stars have planets, and what the statistical distributions of planetary properties look like, we can place our solar system in context and understand which features are typical and which are unusual or rare.

We've also observed stars at various stages of formation, from collapsing clouds just beginning to condense to young stars a few million years old surrounded by protoplanetary disks of gas and dust to slightly older stars where the disk has cleared and planets have presumably formed. These observations show us planetary formation in action, allowing us to test our theories against real systems at different evolutionary stages. Telescopes like ALMA, the Atacama Large Millimeter Array in Chile, can image protoplanetary disks in exquisite detail, revealing gaps and rings and spiral structures that indicate planets are forming within the disk even though the planets themselves are too small and faint to see directly.

Computer simulations represent another powerful tool for understanding Earth's formation. Modern computers can model the gravitational interactions of millions of particles representing dust grains, planetesimals, and planetary embryos, following their orbits and collisions over millions of simulated years to see how planets naturally emerge from disks of debris. These simulations, run thousands of times with slightly different starting conditions, reveal which aspects of planet formation are robust and repeatable versus which depend sensitively on chance events, helping us understand both the typical course of planetary formation and the range of possible outcomes.

Laboratory experiments allow scientists to recreate extreme conditions temperatures of thousands of degrees, pressures of millions of atmospheres that existed during Earth's formation, measuring how materials behave under those conditions. High-pressure experiments using diamond anvil cells, which can squeeze tiny samples between diamonds to generate pressures exceeding those at Earth's core, have revealed how minerals change their structure and properties at depth, informing our understanding of Earth's internal layering. Shock experiments, using high-velocity impacts or explosives to generate the extreme conditions of asteroid collisions, measure how much heating and chemical change occurs during impacts, helping us understand how volatile substances like water might have survived delivery to Earth despite the violence of the accretion process.

Chemical and isotopic analysis of Earth rocks provides clues about processes we can't directly observe. The ratio of different isotopes of elements like oxygen, carbon, nitrogen, and noble gases in Earth materials can be compared to ratios in meteorites, lunar samples, and other solar system bodies to trace where Earth's materials came from and how they've been processed. Tiny variations in these ratios, sometimes just a few parts per million difference, can distinguish between different source regions in the early solar system or different processes that affected the materials.

The laws of physics and chemistry, tested extensively in laboratories and through observations across the universe, provide the framework for understanding what must have happened even when direct evidence is lacking. We know that gravity pulls matter together and that the strength of gravitational attraction increases with mass and decreases with distance. We know that when objects collide at high speeds, kinetic energy is converted to heat. We know that iron is denser than silicate rock and will sink through molten rock under gravity. We know that water has a specific freezing point and boiling point at any given pressure. These laws, combined with knowledge of what materials were present and what energy sources existed, allow us to reconstruct sequences of events with reasonable confidence even when the physical evidence has been destroyed.

From all these sources zircons and meteorites, lunar samples and observations of other stellar systems, computer simulations and laboratory experiments, isotopic measurements and physical laws a coherent picture has emerged of how Earth formed and evolved during its first billion years. It's not a complete picture. Significant uncertainties and active debates remain about many details: exactly when Earth's core finished forming, where Earth's water came from and in what proportions, when plate tectonics started, when and how life originated, whether the Moon-forming impact occurred early or late in Earth's accretion. Different groups of scientists interpret the same evidence differently, propose alternative scenarios, challenge each other's conclusions in the productive dialectic that characterizes healthy scientific inquiry.

But on the major outlines, there is broad consensus. Earth formed approximately 4.54 billion years ago through the accretion of smaller bodies in a disk of material orbiting the young Sun. It was heavily bombarded during and after its formation, experiencing numerous giant impacts including the Moon-forming collision. It differentiated into layers iron core, silicate mantle, thin crust through the sinking of denser materials. It developed an atmosphere through volcanic out gassing. It cooled enough for liquid water to condense and form oceans within perhaps 100 to 200 million years of its formation. Small continental landmasses began forming by 4 billion years ago. Life originated at some point between 4.0 and 3.5 billion years ago and had definitely become established by 3.5 billion years ago.

This book tells that story in chronological order, starting with the cosmic context that made Earth possible and proceeding through planetary formation, differentiation, cooling, the origin of the Moon, the formation of oceans and atmosphere, the emergence of continents, the chemistry that preceded life, and life's first appearance and early evolution. Each chapter builds on previous ones, creating a narrative arc from the Big Bang to the establishment of Earth as a living planet.

The story is told in prose rather than equations, aiming for accessibility to readers without scientific training while maintaining accuracy about what we actually know versus what we're inferring or speculating. When significant uncertainties exist, we'll acknowledge them rather than presenting a false certainty. When the evidence is fragmentary or disputed, we'll say so. The goal is not to give you the illusion of perfect knowledge about events billions of years ago but to show you how scientists have pieced together a coherent narrative from incomplete evidence, and to give you enough understanding of the evidence and reasoning that you could evaluate new discoveries and claims when they arise.

This is fundamentally a work of reconstruction, of detective work, of inference from clues. We cannot go back in time to directly observe Earth's formation. We will never find a fossil of the first living cell or a rock from Earth's original surface. But through careful observation of what has survived, through comparison with other worlds and other systems, through experimentation and simulation and theoretical reasoning, we have learned to read messages written in stone and metal, in isotopes and impact craters, in the distribution of elements and the structures of molecules.

The story these messages tell is more wondrous than any creation myth, more intricate than any fiction, and has the profound advantage of being, to the best of our current knowledge, actually true.

Let us begin.

1

The Universe Learns to Remember

Before there was darkness, there was something we can't name, something that exists outside the vocabulary we've built to describe existence itself. It wasn't nothing nothing is still a concept that needs space to be empty inside of, a void that requires boundaries to define what isn't there. What existed before the universe began was different from nothing in ways that make the question itself nearly meaningless. There was no inside, no outside, no here or there, no up or down. If you'd been present, which you couldn't have been, you wouldn't have seen blackness stretching forever because there was no forever yet, no distance for blackness to stretch across. You wouldn't have seen anything at all, because there was nowhere for light to go, no space for it to travel through, and more importantly, there was no "you" to do the seeing.

Then it started, and everything changed in an instant that was itself the first instant.

Not with a bang, despite what we call it, because there was no air to carry sound waves and no ears to hear them even if there had been. But roughly 13.8 billion years ago, give or take a few hundred million years, everything that would ever exist every star that would ever burn, every planet that would ever form, every thought you've ever had or will have erupted into being from a point far smaller than the period at the end of this sentence. Space itself began expanding, not into anything but rather creating the "anything" as it went, and as it expanded it created time, setting the cosmic clock ticking for the first time. Before this moment, "before" had no meaning because time didn't exist to create a before or an after.

What came next happened fast, at least by the standards of what would come later in the universe's long history. In the first second after the beginning actually, in a fraction of a second so microscopically small it makes a billionth of a second look like an eternity the universe went through changes so violent and strange that our everyday experience offers no comparison, no metaphor that captures even a shadow of what occurred. The temperature was trillions upon trillions of degrees, hotter than anything that exists anywhere in the universe today. Matter and energy were the same thing, constantly switching back and forth, interchangeable in ways that seem impossible now but were normal then. Particles popped into existence from pure energy and then vanished back into it moments later. The fundamental forces that govern everything gravity, electromagnetism, the strong and weak nuclear forces were unified into a single force that would soon split apart into the four distinct forces we know today. The rules of physics as we understand them didn't fully apply yet because the universe was still in the process of deciding what those rules would be, writing its own instruction manual as it expanded.

You've probably heard the phrase that energy can't be created or destroyed, only converted from one form to another, and that's one of the bedrock principles of physics as we know it today. But in that first instant, in that fraction of a second when everything began, the universe seems to have broken its own future rules, pulling all of existence all the energy, all the potential for matter, all of space and time out of something we can't identify or understand. Physicists have developed theories about what might have triggered this beginning, some of them mathematically beautiful, some of them so bizarre they sound like science fiction, involving concepts like quantum fluctuations in a per-existing field or colliding membranes in higher dimensions. None of these theories are certain, and none may ever be proven beyond doubt. The honest answer, the one that requires the most intellectual humility, is that we don't know what happened before time began, and maybe the question doesn't even make sense in the way we're asking it. Asking what came "before" time is like asking what's north of the North Pole the question uses words that stop having meaning at the boundary you're trying to cross.

But while we may never know what sparked the beginning, we do know with considerable confidence what happened next, in the moments and years and millennia that followed.

As the universe expanded, pushing outward in all directions simultaneously, it cooled, and this cooling wasn't incidental but fundamental to everything that would follow. Think of a can of compressed air when you release the pressure and let it expand, the gas gets cold, cold enough to frost over in seconds. It's the same basic principle operating in the universe, just on a scale that makes our atmosphere look like a puddle next to an ocean. The expansion of space itself caused the cosmic temperature to drop, and as things cooled, structure began to emerge from the chaos like crystals forming in a supersaturated solution.

In the first three minutes after the beginning, when the universe was still unimaginably hot hotter than the core of any star that would ever exist protons and neutrons, themselves barely older than the universe and still settling into their identities, began to stick together despite the intense heat trying to blow them apart. They formed the nuclei of the simplest atoms in a process called nucleosynthesis, building the first matter more complex than individual particles. Hydrogen nuclei formed first, just single protons, the simplest possible atomic nucleus. Then helium nuclei formed, each one containing two protons and two neutrons bound together by the strong nuclear force. A tiny amount of lithium formed as well, and trace amounts of a few other light elements, but overwhelmingly it was hydrogen and helium that emerged from this primordial nuclear furnace, in a ratio of about three hydrogen nuclei for every one helium nucleus. That ratio, determined by the temperature and density of the universe in those first minutes, would remain essentially unchanged for billions of years.

That's all the universe managed to make in its infancy, just those two elements dominating everything else. No carbon yet, which would later form the backbone of every living molecule. No oxygen, which would fill the atmospheres of living worlds. No silicon, which would make rocks and sand. No iron, which would sink to the cores of planets and give them magnetic fields. None of the elements that would later make planets possible, let alone life. Just the two simplest atoms, and even they weren't complete atoms yet in the way we'd recognize them today. They were just nuclei, naked cores of protons and neutrons without electrons orbiting them, because it was still far too hot for electrons to settle into stable orbits around anything. The electrons existed, zipping around freely at tremendous speeds, but they were like bees swarming without a hive, too energetic to be captured by the atomic nuclei around them.

These free electrons created a problem for light, or more precisely, they created a problem for the universe's transparency. Photons particles of light couldn't travel freely because they kept crashing into electrons, bouncing off them like pinball in an endless game. Every photon could only travel a tiny distance before colliding with an electron and scattering in a new direction, only to hit another electron almost immediately and scatter again. Light and matter were locked together in an opaque fog.

Picture the thickest fog you've ever seen, the kind where you can barely see your hand in front of your face, where the world disappears just a few feet away from you. Now make that fog a thousand times denser, a million times denser, and fill it with light so intense and so evenly distributed that it would vaporize anything solid instantly. That was the early universe in those first hundreds of thousands of years a glowing, opaque plasma where light couldn't travel more than a microscopic distance before slamming into something else and careening off in a random direction. The universe had tremendous energy, it had matter in the form of hydrogen and helium nuclei and free electrons, but it didn't have clarity or structure. It couldn't see itself yet. Everything was tangled together in a luminous haze.

This state of affairs continued for what seems like a long time by human standards but is remarkably brief in cosmic terms. For 380,000 years a span longer than humans have had written language, longer than we've had agriculture, longer than we've built permanent settlements the universe remained trapped in this hot, dense, opaque condition. If you could somehow have floated in it, protected from the heat by some impossible technology, you would have been completely blind, not because there was no light but because there was too much of it, bouncing chaotically in every direction with nowhere to go, every photon trapped in an endless collision cascade.

Then something remarkable happened, something that would later allow creatures like us to observe and understand the universe's history. The temperature dropped below a critical threshold, crossing a line that changed everything.

It didn't happen in one place before spreading to others it happened everywhere at once, throughout the entire universe, though "everywhere" was much smaller then because the universe was only about one-thousandth its current size. As the temperature fell below roughly 3,000 degrees Kelvin, which is still hot enough to vaporize any material we know but is cool by the standards of the early universe, electrons finally slowed down enough to be captured by atomic nuclei. The electrons, which had been too energetic to be held in orbits, finally settled down into the quantum shells around protons and helium nuclei. Hydrogen atoms became complete structures for the first time: one proton in the nucleus with one electron orbiting it. Helium atoms became complete: two protons and two neutrons in the nucleus, with two electrons in orbit around them.

And just like that, in what must have been a cosmically sudden transition even if it took thousands of years to fully complete, the fog cleared and the universe became transparent.

Photons, which had been trapped in an endless pinball game with free electrons, suddenly found themselves with nothing to collide with because the electrons were now bound up in atoms, hidden inside electron clouds where they were no longer free to scatter light. Light could finally travel freely across space. For the first time since the beginning, the universe became transparent, and photons poured through expanding space that would remain essentially clear for billions of years to come. If you'd been there to witness it, protected somehow from the still-intense radiation and heat, you would have watched the glowing haze fade like morning mist burning off a lake under the rising sun, the luminous fog thinning and vanishing to reveal a vast, dark emptiness beyond.

That light, those photons that were freed when the universe became transparent, is still traveling through space today. We can still detect it, though it's faint now and cold after billions of years of being stretched and red shifted by the universe's ongoing expansion. We call it the cosmic microwave background radiation, and it's the oldest thing we can observe directly—a baby picture of the universe taken when it was less than half a million years old, less than one ten-thousandth of its current age. Every direction we look in the sky, there it is: the afterglow of creation, cooled from thousands of degrees to just 2.7 degrees above absolute zero, but still present, still detectable, still telling the same story about the universe's infancy to anyone with the right instruments to listen.

The universe had learned to remember its own beginning, encoding its history in light that would travel forever.

But here's the strange and somewhat haunting thing about this moment, something worth pausing to appreciate: when the fog cleared and the universe became transparent for the first time, that transformation wasn't witnessed by anyone or anything. No eyes existed anywhere in the cosmos. No minds had formed. Nothing alive existed, nothing even remotely capable of being alive given the conditions. The universe became visible to precisely no one. It was a story unfolding with no audience, a magnificent play performed in a theater that was not just empty but couldn't even have held an audience yet. The drama of existence was occurring in complete solitude.

And yet, the fact that we know about this moment now, that we can reconstruct those early times from mathematics and careful observations made billions of years later, that we can calculate what the temperature must have been and what atoms formed and when the fog cleared that's its own kind of profound achievement. The universe made atoms during that primordial nucleosynthesis, and those atoms eventually, through processes we'll explore in later chapters, made stars, and those stars manufactured heavier atoms in their cores and scattered them through space when they died, and those heavier atoms combined with the original hydrogen and helium to make planets, and on at least one planet orbiting an ordinary star in an ordinary galaxy, some of those atoms arranged themselves into structures complex enough to look back across billions of years and understand where they came from. Matter became conscious of its own history.

You are made of those atoms, and this isn't a metaphor or poetic language but literal physical fact.

2

The Long Darkness

After the flash and fire of the universe's first moments, after the fog cleared and light could finally travel freely through transparent space, you might expect the story to accelerate, to rush forward into the creation of stars and galaxies and all the brilliant structure we see when we look at the night sky today. But that's not what happened. Instead, the universe entered a period so uneventful, so profoundly empty and dark, that astronomers call it the Dark Ages, and the name is entirely appropriate. For hundreds of millions of years a span of time longer than complex life has existed on Earth the universe sat in darkness, waiting for something to happen.

This chapter is about that waiting, about the slow and patient work of gravity operating on timescales that make human history look like a blink. It's about emptiness and potential, about how the universe spent its childhood in the dark, preparing for the moment when it would finally learn to shine.

When the cosmic fog cleared at 380,000 years after the beginning, the universe was left with a nearly uniform distribution of hydrogen and helium gas spread across expanding space. Nearly uniform, but not quite, and that tiny imperfection would make all the difference. The temperature was still around 3,000 degrees Kelvin where atoms had just formed, but it was cooling rapidly as space continued to expand, and within a relatively short time cosmically speaking the universe became genuinely cold and dark. The afterglow from the Big Bang faded from visible light into infrared, then into microwaves, then into radio waves as it was stretched by expansion, becoming dimmer and dimmer until it was essentially imperceptible except as a faint background whisper of radiation.

There were no stars yet to provide light, no galaxies to cluster and shine, nothing but hydrogen and helium atoms drifting through the void, spread so thin that most of space was emptier than the best vacuum we can create in laboratories on Earth today. If you could have floated in this early universe, assuming you could survive without heat or light or anything to breathe, you would have been surrounded by absolute blackness in every direction, a darkness more complete than any cave or ocean depth on Earth could ever achieve. You might have been able to detect the faint microwave glow of the cosmic background radiation if you had the right instruments, but to your eyes if eyes could exist in such conditions there would have been nothing to see at all. No stars above you, no galaxies on the horizon, no light anywhere except that dim background radiation, too faint and too shifted in wavelength for human vision to perceive.

The universe was dark, and it would stay dark for a very long time.

But while nothing was visible, something was happening in that darkness, a process so gradual and subtle that you could have watched for a thousand human lifetimes without seeing any change, yet so consequential that everything in the universe today depends on it. Gravity was working, slowly and inexorably, amplifying the tiny density variations that had been imprinted on the universe during its first moments of existence.

These variations were almost absurdly small, differences in density of less than one part in 100,000, which means that in a region containing 100,000 hydrogen atoms, some spots might have 100,001 atoms while others had

99,999. If the early universe were a perfectly smooth tabletop, these density variations would be like bumps and dips smaller than a single grain of sand on a surface the size of a football field. You could run your hand across the table and never feel the irregularities. They were, by any reasonable measure, negligible.

But gravity doesn't care about reasonable measures, and gravity doesn't ignore small differences. Gravity is patient and persistent, and it responds to mass wherever mass exists, however small the concentration might be. In a region where there were slightly more atoms than average, even just a handful more out of hundreds of thousands, gravity pulled slightly harder than it did in the surrounding space. That extra pull, infinitesimal as it was, drew in more atoms from the nearby regions. And as more atoms accumulated in that spot, the gravitational pull became slightly stronger, which drew in even more atoms, which strengthened the pull further, creating a feedback loop that would eventually transform those microscopic ripples into galaxies and clusters of galaxies.

This is what physicists mean when they talk about gravitational instability, and it's one of the most important processes in the universe. Any region that starts out even slightly denser than its surroundings will, given enough time, become much denser, while the regions around it become emptier as their gas gets pulled away toward the denser areas. The rich get richer, the poor get poorer, and the universe becomes increasingly lumpy and structured rather than remaining smooth and uniform.

But this process doesn't happen quickly, at least not in the early universe when everything was spread so thin. It took time, enormous amounts of time by human standards, for those tiny density variations to grow into structures large enough and dense enough to do anything interesting. The universe had to wait while gravity did its slow, patient work, pulling gas toward the denser regions atom by atom, gradually building up concentrations of matter in some places while leaving vast voids in others.

For perhaps 100 million years after the fog cleared, maybe as long as 200 million years depending on the exact conditions and which region of the universe you're looking at, nothing much changed in any way you could observe. The universe just kept expanding and cooling. The gas kept drifting. Gravity kept pulling, accumulating mass in the denser regions with glacial slowness. The darkness remained absolute and unbroken. No light anywhere except that fading background glow, nothing to mark the passage of time except the steady expansion of space itself and the dropping temperature as the universe grew colder and more dilute.

This must have been the most boring period in cosmic history, if boredom can be applied to a universe that contained no minds to experience it. Nothing burned. Nothing exploded. Nothing changed in any dramatic way. It was just darkness and cold and the slow, invisible work of gravity reshaping the distribution of matter on scales too vast to comprehend, preparing the stage for everything that would follow.

Try to imagine that patience, that cosmic slowness. A hundred million years is longer than the dinosaurs ruled the Earth. It's longer than complex animals have existed anywhere. It's roughly the time it takes for entire mountain ranges to rise and erode away, for continents to drift thousands of miles across the planet's surface. And the universe sat in darkness for that entire span, possibly twice that span, doing nothing visible, making no progress that any observer could have detected, just slowly gathering its resources in certain regions while leaving vast volumes of space nearly empty.

The universe was teaching us something important, though of course there was no "us" yet to learn the lesson. It was demonstrating that the most consequential processes don't always look dramatic while they're happening. Sometimes the most important changes occur invisibly, incrementally, through the patient accumulation of small effects over enormous time scans. The universe wasn't in a hurry. It had billions of years ahead of it, and it was willing to spend millions of those years in darkness, waiting for gravity to build the structures that would eventually become galaxies and stars and planets.

And all the while, in certain favored regions where the density had started out just slightly higher than average, the gas was slowly falling inward, pulled by its own collective gravity. These regions were vast—hundreds of thousands of light-years across, large enough to contain what would eventually become entire galaxies or even clusters of galaxies. The gas within them was still spread incredibly thin by everyday standards, far more diffuse than any vacuum on Earth, but it was beginning to clump together, beginning to separate itself from the cosmic average.

As the gas fell inward toward these denser regions, it began to move, slowly at first but gaining speed as the gravitational pull strengthened. The collapse started to feed on itself. The denser the center became, the harder it pulled on the surrounding gas, and the faster that gas fell inward, and the denser the center became. What had started as an almost imperceptible difference in density a fraction of a percent variation across hundreds of thousands of light-years began to grow into something more substantial.

The gas clouds that were collapsing also began to spin, slowly rotating as they fell inward, and this rotation was crucial for what would come next. In the early universe, when everything was spread relatively uniformly, there were always tiny random motions, small velocities that atoms had in different directions just by chance. As a cloud of gas collapsed under its own gravity, any net rotation, however slight, became amplified. The cloud spun faster as it shrank, the same way an ice skater spins faster when they pull their arms in, conserving angular momentum as the radius decreases. This rotation would eventually flatten the collapsing gas clouds into disks, and within those disks, stars would form.

But we're getting ahead of the story again. During the Dark Ages, these collapsing regions were still enormous and diffuse, still just beginning their long fall toward the densities required to ignite nuclear fusion. They were preparing to become galaxies, gathering the raw material, but they weren't there yet. They were dark clouds of hydrogen and helium, invisible against the dark background of empty space, detectable only by their gravitational influence on the gas around them.

Meanwhile, the universe kept expanding, and this expansion worked against the collapse in an interesting way. While gravity was trying to pull matter together into clumps, the expansion of space itself was trying to pull everything apart, diluting the density and increasing the distances between atoms. In most of the universe, in the regions that had started out with average or below-average density, the expansion won. Those regions became emptier and emptier as their sparse gas was pulled away toward the denser regions, leaving behind vast voids that would eventually become the enormous empty spaces between galaxy clusters. But in the regions that had started out slightly denser than average, gravity won, at least locally. Those regions continued to collapse, fighting against the cosmic expansion, pulling themselves together even as the universe around them tried to pull them apart.

This tension between collapse and expansion, between gravity trying to gather matter and the expansion of space trying to disperse it, is one of the fundamental dynamics that shaped the structure of the universe. If gravity had been slightly weaker, or if the initial density variations had been slightly smaller, the universe might have expanded forever without ever forming stars or galaxies, remaining forever dark and diffuse and empty. If gravity had been slightly stronger, or if the density variations had been larger, everything might have collapsed too quickly, forming massive black holes instead of stars and galaxies. The actual universe found a balance between these extremes, with density variations just large enough, and gravity just strong enough, to create structure without overwhelming it.

The universe was setting up a pattern that would repeat across many scales and many contexts: the balance between forces trying to pull things together and forces trying to push them apart, between stability and change, between order and chaos. This balance would later govern the formation of stars, the structure of galaxies, the orbits of planets, even the chemistry of life itself. But during the Dark Ages, it was playing out on the largest scales, determining where galaxies would form and where vast empty voids would stretch across space.

And still the darkness continued, unbroken by any light source, unchanging to any eye that might have observed it. The universe at 100 million years old looked essentially identical to the universe at 50 million years old or at 150 million years old, at least to any casual observer. The differences were there the collapsing regions were slightly denser, the voids were slightly emptier, the gas was moving slightly faster as it fell inward but these were changes that would only become obvious over millions of years, changes that required patience to detect.

3

Stars That Died So Earth Could Live

Something was about to happen in the darkness, something the universe had never experienced before. In certain dense regions where gas had been falling inward for tens of millions of years, where gravity had compressed hydrogen and helium into clouds thousands of times denser than the cosmic average, the temperature was beginning to rise. Not everywhere, just in specific spots where the collapse had gone furthest, where the gas had packed itself tight enough that atoms were colliding more and more frequently, and each collision generated a tiny amount of heat. Those individual collisions didn't matter much each one raised the temperature by an infinitesimal fraction of a degree but there were trillions upon trillions of collisions happening every second, and over time, over millions of years of continuous collapse, the accumulated heat from all those collisions began to build up in the centers of the densest clouds.

The gas falling inward was doing work in the physics sense of the term, converting gravitational potential energy into kinetic energy and then into heat through friction and collisions. The centers of these collapsing clouds were becoming furnaces, not because anything was burning in the chemical sense there was no oxygen to support combustion, and besides, the temperature wasn't high enough yet but because gravity itself was generating heat through compression, the same way a bicycle pump gets warm when you force air into a tire. The difference was scale. These clouds contained millions or billions of times more mass than our Sun, spread across regions hundreds of light-years wide, and all of that mass was falling toward the center, heating it as it fell.

For a long time, measured in millions of years, the heat could escape. Photons created by the collisions could work their way outward through the cloud, carrying energy away and preventing the temperature from rising too high. The clouds glowed faintly in infrared light, invisible to human eyes but radiating heat into the surrounding darkness. But as the centers grew denser, as more and more mass packed into smaller and smaller volumes, it became harder for photons to escape. They'd get absorbed by an atom, re-emitted in a random direction, absorbed again, bouncing around inside the increasingly opaque gas like light trapped inside a thick fog. The heat started to build up faster, trapped inside the densest cores.

And then, somewhere in the universe we don't know exactly where, we don't know the precise moment, but it happened somewhere around 100 to 200 million years after the Big Bang the temperature in the center of one of these collapsing clouds reached ten million degrees Kelvin.

That number, ten million degrees, is a threshold that changes everything, a line that once crossed opens up entirely new physics. At ten million degrees, atomic nuclei are moving so fast, vibrating so violently, that when they collide they can overcome the electromagnetic force that normally keeps them apart. Protons are positively charged, and positive charges repel each other, which is why atomic nuclei don't normally stick together when they touch they bounce apart like magnets with the same poles facing each other. But at ten million degrees, some of the collisions are so violent that the nuclei smash together hard enough to get past that electromagnetic repulsion and come within range of the strong nuclear force, which is far more powerful than electromagnetism but only operates at extremely short distances, distances smaller than an atomic nucleus.

When two hydrogen nuclei two protons get close enough for the strong force to grab them, something remarkable happens. They fuse together, combining to eventually form helium through a series of reactions, and in the process they release energy, tremendous amounts of energy, far more than the collision that brought them together. This is nuclear fusion, the same process that powers hydrogen bombs, but here it was happening naturally, spontaneously, triggered by nothing but gravity and heat.

That first star, whoever and wherever it was, switched on in the darkness like a lamp in an empty room.

Suddenly there was light where there had been none for hundreds of millions of years. Photons poured out from the star's surface, racing through space at 186,000 miles per second, illuminating the hydrogen gas around it for the first time since the cosmic fog had cleared. The darkness was broken. The universe had learned to shine.

But this first star wasn't alone for long, because the conditions that allowed it to form existed in thousands, perhaps millions of other collapsing clouds scattered across the universe. Within a few million years a cosmic eye blink compared to the hundreds of millions that had passed in darkness stars were igniting throughout the universe, each one a brilliant point of light pushing back against the dark. The universe was experiencing its cosmic dawn, the end of the Dark Ages, the moment when it transformed from a cold, black void into something illuminated and active.

These first stars were nothing like our Sun, nothing like most of the stars we see in the sky today. They were monsters, behemoths, some of them containing a hundred times as much mass as the Sun, some perhaps two hundred or even three hundred times more massive. They burned with a brilliance that would have been blinding to observe, hotter and bluer than any star in the modern universe, and they lived fast because of their enormous size. The more massive a star is, the hotter its core burns and the more quickly it consumes its fuel.

Our Sun will burn steadily for about ten billion years total. These first stars, these Population III stars as astronomers call them, burned through their hydrogen fuel in just a few million years, sometimes less.

They died young, and they died violently, and their deaths were among the most important events in the history of the universe because of what they created in their final moments.

While a star is burning hydrogen in its core, fusing it into helium, the outward pressure from all that fusion energy balances against the inward pull of gravity. The star reaches an equilibrium, a stable size where the two forces exactly counteract each other. But eventually, inevitably, the hydrogen in the core runs out. There's still plenty of hydrogen in the outer layers of the star, but the core where the temperature and pressure are highest, where fusion actually occurs exhausts its supply of protons to fuse together. The fusion reactions slow down and then stop, and suddenly the outward pressure that was holding the star up against its own gravity disappears.

The core collapses.

For a massive star, this collapse is catastrophic and fast, taking just seconds. The core, which had been held up by fusion pressure, suddenly finds nothing supporting it against the enormous weight of all the layers above, and it falls inward, compressing to incredible densities while heating to incredible temperatures. The helium that had been created from hydrogen fusion now finds itself in an environment hot enough to fuse, and it begins combining to form carbon through a process that requires around 100 million degrees. The star stabilizes again, temporarily, burning helium in its core while continuing to burn hydrogen in a shell around the core.

But eventually the helium runs out too, and the core collapses again, and this time it gets hot enough to fuse carbon, creating oxygen, neon, and magnesium. Each stage of fusion creates heavier elements and releases energy that temporarily supports the star, but each stage also burns faster than the last because heavier elements require higher temperatures to fuse and they release less energy when they do. The star is on a countdown now, racing through different fusion stages, creating progressively heavier elements in its core while the outer layers continue burning lighter elements in nested shells, like an onion made of different nuclear reactions.

Carbon fuses into neon and magnesium in thousands of years. Neon fuses into oxygen in about a year. Oxygen fuses into silicon and sulfur in months. And finally, silicon fuses into iron in a matter of days.

Then everything stops, because iron is the end of the line for fusion.

Every element lighter than iron releases energy when it fuses into heavier elements that's what powers the star. But iron is the most stable atomic nucleus that exists, the bottom of the energy curve. Fusing iron into heavier elements doesn't release energy; it requires energy, it costs more than you get back. When a massive star's core turns to iron, it's like a car running out of gas on the highway. The fusion reactions that have been holding the star up for millions of years suddenly have no more fuel to burn. The iron core just sits there, inert and useless, incapable of producing the energy needed to support the star.

And so the core collapses one final time, but this collapse is different from all the previous ones because there's no fusion reaction waiting to save it, no new fuel to ignite. The iron core collapses all the way down, compressing until the electrons are pushed into the atomic nuclei, combining with protons to form neutrons in a process that removes the electron pressure that had been providing the last bit of support. The core becomes a ball of neutrons, packed so tightly that a teaspoon of the material would weigh as much as a mountain, compressed to the point where it can't collapse any further because the neutrons themselves resist being pushed closer together through quantum mechanical effects that prevent them from occupying the same space.

The collapse happens in less than a second, and when the infilling core hits that neutron-matter density limit, it bounces.

The bounce sends a shock wave exploding outward through the star at a significant fraction of the speed of light, and this shock wave tears the star apart from the inside out. The outer layers, which had been falling inward following the collapsing core, suddenly encounter this expanding wall of energy and get blasted back outward with tremendous force. The star explodes in what we call a supernova, one of the most violent events in the universe, releasing in a few seconds more energy than the star had produced during its entire previous lifetime. For a brief period days or weeks a single exploding star can shine as brightly as an entire galaxy containing hundreds of billions of normal stars.

But the energy release, spectacular as it is, isn't the most important part of what happens in a supernova, at least not for our story. The most important part is what the explosion does to the elements inside the star.

During the normal lifetime of a star, fusion only creates elements up to iron because that's where fusion stops being energy-positive. But in the extreme conditions of a supernova temperatures of billions of degrees, pressures beyond anything that exists in a normal star, floods of neutrons released from the collapsing core nuclei can be forced to fuse past iron, creating all the heavier elements in the periodic table. Gold, silver, platinum, uranium, lead every element heavier than iron was created in supernova explosions or in related violent events like neutron star collisions. The explosion doesn't just create these elements; it scatters them into space, blasting the star's material outward at thousands of miles per second, seeding the surrounding gas with heavy elements that never existed there before.

Those first stars, those massive Population III giants that formed in the early universe, were element factories, and their deaths enriched the cosmos with the raw materials needed to build something more complex than just hydrogen and helium gas.

Before the first stars, the universe contained essentially only hydrogen and helium, plus a trace of lithium. After the first supernovae, the universe contained carbon, oxygen, nitrogen, silicon, iron, and dozens of other elements, scattered through the gas clouds in tiny concentrations but present for the first time. The universe had been upgraded, chemically speaking, given a periodic table that contained more than just the first two entries.

And this matters for Earth it matters fundamentally and absolutely because you cannot build a planet like Earth out of hydrogen and helium alone. Hydrogen and helium are gases at any temperature you're likely to find in space near a star. They don't condense into solids unless you compress them with enormous gravity, which is why the gas giant planets like Jupiter are so massive and why they're found in the outer solar system where it's cold. To build a rocky planet, a planet with a solid surface and an iron core and oceans of liquid water, you need heavier elements. You need oxygen to combine with hydrogen and make water. You need silicon and oxygen to make the silicate minerals that form rock. You need iron to sink to the center and form a core. You need carbon and nitrogen and phosphorus for life.

Every one of those elements was created inside a star.

The carbon in every living cell on Earth, the oxygen you're breathing right now, the calcium in your bones, the iron in your blood all of it was forged inside stars that exploded billions of years ago, long before the Sun existed, long before the Earth had even begun to form. This isn't poetic exaggeration or loose metaphorical language but verified scientific fact, as certain as anything we know about the universe. Every atom in your body heavier than hydrogen was manufactured in the nuclear furnace at the core of a star, and most of those atoms were scattered into space when that star died violently, exploding with enough force to outshine entire galaxies for a brief moment.

The first generation of stars lived fast and died young, burning through their fuel in just a few million years before exploding as supernovae, and when they exploded they polluted the surrounding gas with heavy elements. That word "polluted" might seem strange we usually think of pollution as contamination, as making something worse but from the perspective of building planets and eventually life, this pollution was essential enrichment. The pristine hydrogen and helium gas of the early universe was too simple, too chemically limited to make anything interesting. It needed to be contaminated with heavier elements, seeded with the periodic table, before the universe could build worlds like Earth.

After the first supernovae, the gas clouds floating through space were no longer pure hydrogen and helium. They contained perhaps one or two percent heavier elements what astronomers call metals, even though the category includes nonmetals like oxygen and carbon mixed throughout. This might not sound like much, one or two percent contamination, but it was enough to change everything about how the next generation of stars would form and what would happen around them.

When the second generation of stars began to condense from these enriched gas clouds, perhaps a few hundred million years after the first stars had ignited, they formed differently than their predecessors had. The heavy elements, even in small concentrations, changed the physics of the collapsing clouds. Carbon and oxygen atoms are much better at radiating away heat than hydrogen and helium are, more efficient at converting thermal energy into photons that can escape from the cloud. This meant that clouds containing heavy elements could cool more effectively as they collapsed, and cooler clouds fragment more easily into smaller pieces rather than forming single massive objects.

The second generation of stars tended to be smaller than the first generation, more like the stars we see today, with a wider range of masses from tiny red dwarfs barely massive enough to sustain fusion up to large blue giants similar to the first stars but not quite as extreme. Our Sun is probably a third or fourth generation star, formed from gas that had been enriched by multiple rounds of stellar death and rebirth, containing about two percent elements heavier than helium. That two percent makes all the difference between a solar system like ours, with rocky planets and the possibility of life, and a solar system that contains only gas giants or perhaps no planets at all.

But we're getting ahead of ourselves again, jumping billions of years into the future when we should still be in the early universe, watching the first few generations of stars live and die and slowly, gradually enrich the cosmos with the elements needed for planets.

The cycle of stellar birth and death became the fundamental rhythm of the universe's chemical evolution. Stars formed from gas clouds, burned for millions or billions of years depending on their mass, created heavier elements in their cores through fusion, and then returned those elements to space either through gentle stellar winds that blow off the outer layers or through violent supernova explosions that scatter everything. The enriched gas would later collapse again to form new stars, which would create even more heavy elements and scatter them in turn, continuing the cycle.

Each generation of stars started with slightly more heavy elements than the previous generation, building up the cosmic abundance of carbon and oxygen and silicon and iron over billions of years. It's a slow process, this chemical enrichment, because most of the mass in any gas cloud is still hydrogen and helium the heavy elements remain trace contaminants even after billions of years of stellar processing. But those trace amounts are enough. Two percent heavy elements can build entire planetary systems, entire worlds, entire biospheres.

The massive stars of the first few generations were doing more than just creating elements, though that alone would have been remarkable enough. They were also shaping their environments through the sheer force of their existence. A massive star doesn't sit quietly in space burning its fuel it blasts its surroundings with intense radiation and powerful stellar winds, streams of charged particles flowing outward at hundreds of miles per second. This radiation and wind push against the surrounding gas, carving out bubbles and cavities, compressing some regions and dispersing others, triggering new waves of star formation in the compressed areas while preventing it in the dispersed ones.

When these massive stars exploded as supernovae, the impact on their surroundings was even more dramatic. The expanding debris from the explosion plowed into the surrounding gas at thousands of miles per second, creating shock waves that compressed the gas and triggered new star formation. Supernova remnants are among the most beautiful objects in the modern universe when we observe them with telescopes vast shells of glowing gas expanding through space, intricate filaments and structures lit up by the energy of the explosion and the radioactive decay of elements created in the blast. But in the early universe, these explosions were doing more than creating pretty structures; they were shaping the formation of galaxies themselves, determining where stars would form and where gas would be blown away into the intergalactic void.

The first few hundred million years after the cosmic dawn were chaotic and violent by the standards of the modern universe. Stars were forming rapidly in certain dense regions, burning brightly, exploding as supernovae, triggering more star formation with their shock waves, creating a cascade of stellar birth and death that lit up the darkness with increasing brilliance. The universe was learning what it could do with matter and energy, experimenting with different sizes and types of stars, building up the chemical complexity that would later allow for planets and life.

And gradually, over hundreds of millions of years, the randomly scattered sites of star formation began to organize themselves into larger structures. Gravity was still working on the largest scales, pulling together not just individual gas clouds but entire regions containing millions or billions of solar masses of material. These regions were becoming the first galaxies, collections of millions of stars bound together by their mutual gravity, swirling through space as coherent objects rather than just random scatterings of light.

The early galaxies were smaller and more irregular than the majestic spiral and elliptical galaxies we see today, more like clumps and blobs of stars than organized structures, but they represented a new level of organization in the universe. Where before there had been just gas and individual stars, now there were gravitation ally bound systems containing millions of stars, systems that could persist for billions of years, that could continue forming new stars from their internal reservoirs of gas, that could collide and merge with each other to build even larger galaxies.

Our own Milky Way galaxy formed from the merger of many smaller galaxies during this early period, accumulating mass and stars over billions of years, though we're again jumping ahead in our story. The point is that the universe was structuring itself on multiple scales simultaneously atoms within stars, stars within galaxies, galaxies within clusters of galaxies building up a hierarchy of organization from the nearly uniform distribution of gas that had existed at the end of the Dark Ages.

And all of this activity, all of this stellar fusion and supernova violence and galaxy formation, was ultimately driven by gravity working on the tiny density fluctuations that had been imprinted on the universe during its first instant of existence. Those microscopic ripples in the density of the cosmic gas, barely one part in 100,000 different from the average, had grown over hundreds of millions of years into structures containing trillions of times more mass than our Sun, into galaxies and clusters that would persist for the rest of cosmic history.

But let's return to the elements themselves, to the specific atoms that would billions of years later come together to form Earth, because that's the thread we need to follow if we want to understand our planet's beginning.

Imagine a carbon atom created in the core of a massive star about twelve billion years ago, somewhere in a region of space that would eventually become part of our galaxy. This carbon atom started as three helium nuclei that collided and fused together in a process that requires extraordinarily precise conditions the temperature has to be around 100 million degrees, and the density has to be high enough that three helium nuclei can find each other and stick together before they drift apart. The process is called the triple-alpha reaction because helium nuclei are also called alpha particles, and it's the only way to make carbon in significant quantities.

Our hypothetical carbon atom existed in the star's core for perhaps a few thousand years, just one carbon nucleus among billions of trillions, surrounded by a sea of helium that was still being converted into carbon and oxygen through ongoing fusion. The temperature was so high that atoms as we normally think of them didn't exist the carbon nucleus had no electrons orbiting it, couldn't form chemical bonds, was just a bare nucleus of six protons and six neutrons held together by the strong nuclear force.

Then the star's core ran out of helium to fuse, and the core collapsed, and the temperature rose high enough to start fusing carbon into heavier elements like neon and magnesium. But our specific carbon atom happened to be located in a region of the core that didn't quite reach the temperatures needed for carbon fusion, or maybe it was in an outer layer where the conditions weren't extreme enough. By chance, by accident, it survived the next round of fusion without being transformed into something else.

When the star finally exploded as a supernova, our carbon atom was blown into space along with everything else, accelerated to thousands of miles per second by the force of the blast, heated to millions of degrees by the shock wave, finally able to capture electrons and become a complete carbon atom rather than just a bare nucleus. It spent the next several billion years drifting through the interstellar medium, the thin gas between stars, occasionally colliding with other atoms but mostly just floating through the vast emptiness, gradually cooling, gradually slowing down as it interacted with the surrounding gas.

It might have collided with oxygen atoms and formed carbon monoxide molecules, which are common in interstellar space. It might have stuck to the surface of a tiny dust grain made of silicates and ice, becoming part of a particle the size of a smoke particle from a cigarette. It definitely traveled enormous distances hundreds or thousands of light-years as it drifted through the galaxy, carried by the gentle currents of gas flowing between stars, pushed around by stellar winds and supernova shock waves, participating in the great recycling system that moves matter from stars back into the interstellar medium and eventually into new stars.

4

A Cloud Collapses

About four and a half billion years ago, give or take a few tens of millions of years, something happened in an unremarkable region of space roughly 26,000 light-years from the center of the Milky Way galaxy. There was nothing special about this location, nothing that marked it as historically significant or cosmically important. It sat in what astronomers call the Orion Arm, a minor spiral arm between two larger structures, in a neighborhood that contained perhaps a few hundred thousand stars scattered across hundreds of light-years. The region had witnessed billions of years of stellar evolution by this point generations of stars forming, living, dying, enriching the local gas with heavy elements through their deaths. The gas floating through this part of the galaxy contained about two percent elements heavier than helium, a typical metallicity for our region of the Milky Way at that time, the accumulated product of billions of years of stellar nucleosynthesis.

And there was a cloud, an enormous diffuse collection of gas and dust that had been drifting through this region for perhaps tens of millions of years, possibly longer. This wasn't a cloud in the sense we usually think of clouds on Earth fluffy, opaque, with defined edges and obvious boundaries. It was far more tenuous than that, a region where the gas was slightly denser than the average interstellar medium, perhaps a few hundred or a few thousand atoms per cubic centimeter compared to roughly one atom per cubic centimeter in the surrounding space. By the standards of any vacuum we can create on Earth, where even the best vacuum chambers still contain billions of atoms per cubic centimeter, this cloud was essentially empty space. But by the standards of interstellar gas, it was relatively dense, and more importantly, it was large, stretching across perhaps a hundred light-years in its longest dimension and containing enough mass to make thousands of stars the size of our Sun.

The cloud was cold, just a few tens of degrees above absolute zero, kept cool by its ability to radiate heat into space through infrared light emitted by molecules and dust grains. It was also dark, completely invisible to human eyes because it emitted no visible light and because the tiny dust particles mixed throughout the gas blocked any visible light trying to pass through from background stars. If you could have floated inside this cloud, protected somehow from the cold and the vacuum, you would have been surrounded by absolute blackness, unable to see your hand in front of your face even though you'd technically be in outer space where you might expect to see stars in every direction. The dust would have blocked everything, creating a darkness more complete than any cave on Earth.

This cloud had been stable for millions of years, maintaining roughly the same size and density, neither collapsing under its own gravity nor dispersing into the surrounding space. It was balanced between two opposing tendencies gravity pulling the gas inward, trying to compress the cloud, and thermal pressure pushing outward, trying to expand it. The gas had some random internal motions, currents and eddies stirring it gently, and it was rotating very slowly as a whole, completing perhaps one rotation every few million years. These motions created turbulence that provided additional pressure to resist gravitational collapse, holding the cloud up against its own weight.

But this balance was fragile, dependent on conditions remaining stable, and conditions in space are never truly stable for very long.

Something disturbed the cloud, triggered it to begin collapsing, and we can't know with certainty what that trigger was because the evidence has long since vanished. Perhaps a nearby massive star exploded as a supernova, and the shock wave from that explosion crashed into the cloud, compressing one edge of it enough to overcome the internal pressure that had been holding it up. Perhaps a smaller cloud drifting through space collided with our cloud, adding mass in one region and tipping the balance toward collapse. Perhaps the cloud simply accumulated enough mass through gradual accretion from the surrounding medium that it eventually became gravitation ally unstable without any dramatic external trigger, reaching a point where its own gravity could overcome the internal pressure and turbulence.

Whatever the cause, sometime around four and a half billion years ago, the cloud began to fall inward, and once that collapse started, it accelerated and spread, pulling in more and more of the cloud's mass in a gravitational avalanche that would reshape the entire region.

The collapse didn't happen uniformly across the whole cloud because the cloud wasn't uniform to begin with. Some regions were slightly denser than others, some slightly warmer, some moving in slightly different directions. As the cloud began to fall inward, these irregularities became amplified. The denser regions collapsed faster because their stronger gravity pulled harder, and as they collapsed they became even denser, pulling even harder, creating local centers of collapse within the larger collapse. The cloud fragmented as it fell, breaking into many smaller clouds, each one collapsing under its own gravity, each one potentially destined to become a star or a small cluster of stars.

One of these fragments not the largest, not the smallest, but somewhere in the middle of the size distribution would eventually become our solar system, containing the Sun and Earth and all the other planets. Let's follow that particular fragment as it separated from its siblings and began its own journey toward becoming something new.

This proto-solar cloud, as we can call it, contained roughly the mass of the

Sun plus a bit extra for planets and other objects, about two percent more massive than the Sun is today. Most of that mass was still hydrogen and helium in roughly the same proportions that the Big Bang had created about 74 percent hydrogen by mass, about 24 percent helium, and only about 2 percent everything else. But that 2 percent was crucial because it contained all the elements needed to build planets: oxygen, carbon, nitrogen, silicon, magnesium, iron, sulfur, calcium, aluminum, and dozens of others in smaller amounts.

The heavy elements weren't evenly distributed within the gas. Some of them existed as individual atoms mixed in with the hydrogen and helium, but many had already combined into molecules or had condensed onto the surfaces of tiny dust grains, solid particles smaller than the particles in cigarette smoke, each grain perhaps a tenth of a micron across. These dust grains were incredibly important for what would happen next, even though they represented less than one percent of the cloud's total mass, because they would serve as the seeds around which planets would eventually form.

As the proto-solar cloud collapsed, it began to spin faster, much faster, accelerating its rotation in accordance with the conservation of angular momentum. Think of an ice skater pulling their arms in during a spin the same physics applies to a collapsing cloud of gas and dust. The cloud had been rotating very slowly to begin with, perhaps taking millions of years to complete one rotation, but as it shrank, the rotation accelerated. By the time the cloud had collapsed to a few light-years across, it was rotating noticeably, completing a rotation in tens of thousands of years. By the time it had collapsed further to a few trillion miles across, it was spinning once every few thousand years.

This rotation had profound consequences for the cloud's shape and structure. If the cloud had been rotating slowly or not at all, it would have collapsed more or less spherically, falling inward equally from all directions toward the center. But because it was rotating, material falling inward from the poles the directions parallel to the rotation axis could fall directly toward the center without interference. Material falling inward from the equatorial regions, however, couldn't fall straight in because it was moving sideways as well as inward, carrying angular momentum that prevented it from reaching the center. Instead, it settled into a rotating disk perpendicular to the rotation axis, a flattened structure where the centrifugal force from rotation balanced against the gravitational pull toward the center.

The cloud transformed from a roughly spherical blob into a disk with a central bulge, somewhat like a pizza with a thick crust around a dense center, though that comparison understates the scale by many orders of magnitude. The disk extended billions of miles across, perhaps even tens of billions of miles, while the central bulge was much smaller, maybe tens of millions of miles in diameter, concentrating most of the cloud's mass in a relatively tiny volume at the center.

This structure a dense central condensation surrounded by a rotating disk is called a protoplanetary disk or sometimes a protostellar disk, and it's the standard configuration that forms whenever a cloud collapses to form a star. We see them around young stars throughout the galaxy today, beautiful structures when observed with modern telescopes, though we can only see the ones that are close enough and oriented favorably relative to Earth. The disk around the proto-Sun looked essentially similar to the disks we observe around other young stars, made of gas and dust swirling around a central concentration that was growing hotter and denser as more material fell onto it.

The temperature within this disk varied enormously depending on distance from the center, creating different zones with radically different conditions. Close to the forming Sun, within a few million miles of the center, the temperature was blazingly hot thousands of degrees heated by the gravitational energy being released as gas fell inward and by the growing radiation from the central condensation. At these temperatures, most materials evaporated.

Only the most refractory substances materials with extremely high melting points like certain metals and metal oxides could exist as solids. Everything else was vaporized, existing as individual atoms or simple molecules in the gas phase.

Farther out from the center, perhaps around fifty to a hundred million miles away roughly the distance where Earth would eventually form the temperature was cooler, maybe a few hundred degrees to perhaps a thousand degrees depending on the exact distance and the stage of collapse. At these temperatures, rocks could condense. Silicate minerals compounds of silicon, oxygen, and various metals like magnesium, iron, and aluminum could form solid grains. Iron and nickel metal could exist as solid particles. These substances have high melting points, considerably higher than water or methane or ammonia, so they could condense into solids relatively close to the forming Sun where things were still quite hot by everyday standards.

Beyond about three or four times Earth's eventual distance from the Sun, past a boundary that astronomers call the frost line or snow line, the temperature dropped below the freezing point of water, and suddenly there was much more solid material available. Water ice could condense from water vapor, and water was by far the most abundant compound made from heavy elements oxygen was the third most common element in the universe after hydrogen and helium, and most of that oxygen had combined with hydrogen to form water molecules. Similarly, methane ice and ammonia ice could condense in the coldest outer regions. These ices dramatically increased the amount of solid material available in the outer disk compared to the inner regions, perhaps by a factor of three or four, because water and other volatile were so much more abundant than rocky materials.

This temperature gradient, this variation in conditions from the hot inner disk to the cold outer disk, would determine what kinds of planets formed where. The inner regions, where only rocks and metals could condense, would produce small rocky planets with iron cores Mercury, Venus, Earth, and Mars. The outer regions, beyond the frost line where ices could condense, would produce much larger planets that could accumulate enormous atmospheres of hydrogen and helium Jupiter, Saturn, Uranus, and Neptune. The architecture of the solar system, the basic layout of what types of planets exist at what distances, was being determined by temperature and chemistry in the protoplanetary disk, long before the planets themselves had begun to form.

But temperature wasn't the only variable that mattered. The disk was also turbulent, full of swirling eddies and currents like water flowing in a river, driven by the complex interactions between gravity, rotation, pressure, and magnetic fields. Gas and dust weren't simply orbiting peacefully around the forming Sun in perfect circles; they were stirred and mixed by turbulence that created complex three-dimensional flows. Near the mid plane of the disk the thin layer closest to the disk's equator where most of the dust concentrated the turbulence was relatively gentle, allowing dust particles to settle and accumulate. But higher above and below the mid plane, the turbulence was stronger, keeping smaller particles suspended and preventing them from settling.

The dust grains floating in the disk were constantly colliding with each other, bumping together as they orbited the Sun at slightly different speeds and altitudes. Most of these collisions were gentle enough that the grains stuck together rather than bouncing apart or shattering, held by weak forces like static electricity and van der Waals forces the same molecular attractions that allow a gecko to climb a wall. Two grains would collide and stick, forming a slightly larger grain. That larger grain would collide with another grain and stick, growing larger still. Slowly, gradually, over thousands of years, the dust grains grew from sub micron particles into larger aggregates the size of sand grains, then pebbles, then rocks.

This process of grains sticking together through collisions is called accretion, and it's the fundamental mechanism by which planets form, though the process works differently at different scales. At the dust grain scale, grains stick together because of molecular forces. Once objects grow to the size of small asteroids, gravity becomes the dominant force holding them together and pulling in more material. But there's an awkward intermediate stage, when objects are somewhere between a boulder and a small mountain in size, where neither molecular forces nor gravity works particularly well, and this is where planet formation faces its biggest challenge.

Objects the size of boulders or small buildings are too large to stick together easily through molecular forces they're likely to bounce apart or shatter when they collide at the speeds typical in the protoplanetary disk, which could be several miles per second. But they're also too small to have significant gravitational pull to attract and hold onto other objects. This is called the meter-scale barrier or the bouncing barrier, and it represents a major obstacle in planet formation theory. How do you get from pebble-sized objects that stick together easily to mountain-sized or asteroid-sized objects that have enough gravity to grow by attracting more material, when the intermediate sizes seem unlikely to stick together at all?

Astronomers have proposed several solutions to this problem, and the real answer probably involves a combination of mechanisms. In some regions of the disk, turbulence might concentrate dust into local dense patches where particles collide frequently enough that some grow large despite the bouncing barrier. In other regions, objects might grow rapidly through a process called pebble accretion, where small pebbles spiral inward through the disk due to gas drag and get swept up by slightly larger objects, allowing those objects to grow quickly without requiring direct collisions between large bodies. There might also be locations in the disk where conditions allow objects to stick together more easily regions where ice coatings make surfaces stickier, or where lower collision speeds reduce the violence of impacts.

Whatever the exact mechanisms, somehow the disk managed to overcome the meter-scale barrier, because we know that larger objects did form. Within perhaps a million years after the disk formed, there were objects scattered throughout it that had grown to the size of small asteroids a mile across, ten miles across, perhaps a hundred miles across in some cases. These objects, called planetesimals, were large enough to have measurable gravity, enough to gravitation ally attract smaller objects nearby and sweep them up, adding to their mass.

The planetesimals didn't form uniformly throughout the disk. They formed more readily at certain distances from the Sun where conditions were favorable perhaps where the disk was denser, or where ice lines created local accumulations of solid material, or where turbulence created persistent pressure bumps that trapped dust. The disk probably had rings or bands where planetesimals were more common, separated by gaps where they were rarer, giving it a structure somewhat like a cosmic parking lot with cars concentrated in certain rows.

Once planetesimals existed, the pace of growth accelerated dramatically because gravity is a runaway process—the more massive an object becomes, the stronger its gravitational pull, and the faster it can accumulate more mass, which makes it even more massive, which strengthens its pull further. A planetesimal ten miles across might take a million years to form through the slow accretion of smaller particles. But once it existed, it could double its mass by attracting nearby planetesimals in perhaps a few hundred thousand years. And that doubled object could double again in less time, growing exponentially rather than linearly.

The disk was becoming a violent place, a shooting gallery where planetesimals crashed into each other constantly, colliding at speeds of several miles per second, often shattering both objects but sometimes allowing the larger one to capture fragments from the smaller one. Most collisions were destructive at this stage, breaking objects apart rather than building them up, but enough collisions resulted in net growth that the process continued. The largest objects in each region of the disk grew faster than their neighbors, pulling in more and more material, gradually dominating their zones.

Meanwhile, at the center of the disk, the proto-Sun was continuing its own collapse, growing hotter and denser as more gas fell onto it from the surrounding disk. The central condensation now contained perhaps 99 percent of the disk's total mass, having accredited most of the material that had fallen inward. It was glowing now, not from nuclear fusion yet but from gravitational energy, radiating infrared light as the gas compressed and heated. The surface temperature might have been a few thousand degrees, hot enough to glow dull red if anyone had been there to see it, though the light would have been hidden by the surrounding disk and the dense cloud of gas and dust that still enveloped the whole system.

The proto-Sun was not yet a star in the true sense because nuclear fusion had not ignited in its core the core wasn't hot enough yet, still below the ten million degree threshold needed for hydrogen fusion. But it was on its way, growing hotter as it continued to contract, perhaps just a few million years away from reaching the temperatures and densities required to start burning hydrogen. For now it was a protostar, a star in the process of formation, powered entirely by gravity rather than fusion.

The protostar was also generating powerful magnetic fields, threaded through the rotating disk and connected to the star's surface, and these magnetic fields were channeling material in complex ways. Gas falling toward the star from the disk didn't fall directly onto the surface but was diverted by magnetic field lines, funneled toward the magnetic poles where it crashed onto the star in hot spots that blazed with X-ray emission. The magnetic fields also launched powerful winds streams of charged particles flowing outward along field lines, carrying away angular momentum and helping to slow down the star's rotation, which otherwise would have spun up to dangerous speeds as the star contracted.

Some of the out flowing wind material was focused into narrow jets shooting outward along the rotation axis, perpendicular to the disk, moving at hundreds of miles per second. These jets are a common feature of young stars, visible in observations of star-forming regions today, and they represent a way for the system to dispose of excess angular momentum and energy. The jets carved through the surrounding gas cloud, creating cleared paths that allowed light from the protostar to escape, though the disk itself still blocked most of the light from reaching the regions where planets would form.

The disk around the proto-Sun was also evolving, losing mass over time as material either fell inward onto the star or was blown away by stellar winds and radiation. The gas in the disk didn't orbit in perfect circles but spiraled slowly inward, pulled toward the star by friction with other gas and by the removal of angular momentum through magnetic fields. This inward drift meant that the disk was temporary, existing for perhaps only a few million years before most of its gas would be either accredited onto the star or dispersed into space. The planets would have to form quickly, growing from dust and planetesimals into full-sized planets before the gas disappeared, because once the gas was gone, many of the processes that enabled rapid growth would stop.

5

The Violent Birth of Earth

Earth did not form gently or peacefully, and it certainly didn't assemble itself according to any plan or design. It was built through catastrophic violence, through millions of collisions between objects ranging from dust grains to small planets, through impacts so destructive that they vaporized rock and melted entire worlds, through a process that was fundamentally chaotic and random yet somehow produced a planet capable of hosting oceans and life. The story of Earth's formation is a story of cosmic demolition, of matter being smashed together repeatedly until gravity finally managed to organize it into something stable.

The region of the protoplanetary disk where Earth would eventually form roughly 93 million miles from the forming Sun, though the exact distance varied as orbits shifted over time started out as nothing more than dust and gas swirling around the central protostar. This was within the region where temperatures allowed rocky materials to condense but were too hot for water ice to exist as a solid, so the solid material available consisted mainly of silicate minerals and metal grains, primarily iron and nickel. There was much less solid material here by mass than in the outer solar system beyond the frost line, perhaps only about half a percent of the total mass in this region existed as solids, with the rest being hydrogen and helium gas that would eventually either fall into the Sun or be blown away by stellar winds.

But half a percent of the mass in a region extending billions of miles was still an enormous amount of material, enough to build several Earth-sized planets if it could all be gathered together. The challenge was gathering it, organizing it from a thin distribution of dust spread across vast distances into concentrated bodies with enough mass to hold themselves together gravitation ally.

The first phase of Earth's formation, the growth from dust grains to planetesimals perhaps a mile or ten miles across, probably took somewhere between a few hundred thousand years and a million years, though the exact timeline is uncertain and varied at different locations in the disk. This phase proceeded through the gentle accretion processes we discussed in the previous chapter grains sticking together, building larger aggregates, somehow crossing the difficult meter-scale barrier to become objects large enough to have meaningful gravity. By the time this phase completed, the region around Earth's future orbit contained perhaps millions of planetesimals, rocky objects ranging from the size of a large building to the size of a small mountain, orbiting the Sun in slightly different paths, occasionally crossing each other's orbits and colliding.

The second phase, the growth from planetesimals to planetary embryos perhaps a thousand miles across, was faster and more violent. This phase was dominated by gravitational interactions and collisions between planetesimals, and it probably took somewhere between 100,000 years and a few million years depending on local conditions. The process worked through runaway accretion the largest objects in any region grew faster than their neighbors because their stronger gravity allowed them to sweep up material from a larger volume of space. A planetesimal ten miles across might have a gravitational reach extending a few tens of miles around it, capturing any smaller object that passed within that distance. But a planetesimal a hundred miles across had a much stronger pull, reaching out hundreds of miles, and could dominate a much larger feeding zone.

As the largest objects grew, they began to clear their orbital zones, sweeping up most of the material in their vicinity, leaving the disk increasingly empty between the surviving large bodies. The number of objects decreased while their individual masses increased. Instead of millions of small planetesimals, the inner solar system now contained perhaps a few dozen to a few hundred planetary embryos, objects ranging from Moon-sized to Mars-sized, each one orbiting the Sun in a path that occasionally brought it close to one or more of its neighbors.

These embryos were already substantial worlds, large enough to be nearly spherical under their own gravity, large enough to have differentiated interiors if they'd been heated sufficiently meaning their denser materials like iron had begun to sink toward the center while lighter materials like silicate rocks floated toward the surface. Some of them might have had thin atmospheres of gases released from their interiors or captured from the surrounding nebular gas. They were proto-planets, not quite full planets yet but heading in that direction, needing only more mass and more time to reach planetary status.

And then they began to destroy each other.

The orbits of planetary embryos are never perfectly stable over long timescales when multiple large bodies exist in nearby orbits. Gravitational interactions between the embryos gradually changed their orbits, pulling them slightly this way or that each time they passed relatively close to each other. Over millions of years, these small perturbations accumulated, stretching orbits into more elliptical shapes, tilting them relative to each other, creating situations where bodies that had once been on well-separated paths now crossed each other's orbits regularly. The system became gravitation ally unstable, and instability led inevitably to collisions.

When two planetary embryos collided, both traveling at several miles per second in orbits around the Sun, the impact released more energy than all the nuclear weapons humans have ever built, enough energy to vaporize substantial portions of both objects and melt everything else. The collision would have been instantaneous from a human perspective the two bodies approaching each other, making contact, and then suddenly exploding in a spray of vaporized rock and molten debris, all in a few seconds or perhaps a few minutes at most. The material from both bodies would have mixed together, some of it escaping into space to orbit independently, some of it falling back together under mutual gravity to form a single larger body, still partially molten from the heat of the collision.

This was how planetary embryos grew into planets, through a process astronomers call giant impacts or late-stage accretion, and it was fundamentally different from the earlier gentle sticking-together of dust grains or the relatively orderly sweep-up of planetesimals. This was violent, destructive, and chaotic. Most of the mass that eventually became Earth probably came from perhaps ten to twenty giant impacts between embryos, each impact adding roughly five to ten percent of Earth's final mass, each one leaving the growing planet molten and glowing, taking millions of years to cool down before the next major collision occurred.

But there were also thousands of smaller impacts during this period, collisions with planetesimals and minor embryos that delivered material more gradually, adding perhaps half of Earth's mass through a continuous rain of smaller bodies. These impacts were less dramatic individually but collectively delivered enormous amounts of energy, keeping Earth's surface hot and molten throughout the accretion period. Every impact, regardless of size, converted the kinetic energy of the incoming body into heat the faster the object was moving and the more massive it was, the more heat it generated on impact. And objects colliding with a growing planet are moving fast, accelerated by the planet's gravity as they fall inward, typically hitting at several miles per second even if their initial approach velocity was relatively low.

Earth grew in fits and starts over a period that probably lasted between ten and a hundred million years the uncertainty is large because we don't have precise measurements of how fast this process occurred, only broad constraints from computer simulations and the radioactive dating of the oldest materials in the solar system. The growth was not smooth or steady but episodic, with long quiet periods when relatively little material accumulated, interrupted by sudden giant impacts that added large amounts of mass in a single event.

During this time, Earth was nothing like the planet we know today. It had no solid surface in the conventional sense, no crust of solid rock that you could stand on, assuming you could survive the heat and the lack of atmosphere and the constant bombardment. The entire planet was largely or entirely molten, a ball of liquid rock thousands of miles in diameter, glowing orange or red from the heat of formation, radiating away energy into space through thermal radiation. The surface temperature might have been 1,500 to 2,000 degrees Celsius, hot enough to melt any rock, hot enough that the surface would have glowed visibly, bright enough to read by if you were somehow hovering above it in a protected spacecraft, though of course no one was there to witness it.

The molten surface was not calm or static but actively convicting, with hot material rising from deeper levels and cooler material sinking back down in enormous circulation patterns, similar to boiling water but on a planetary scale and moving much more slowly because molten rock is extremely viscous, thousands of times thicker than water even when completely liquid. These convection currents created patterns on the surface regions of up welling where fresh hot material emerged, regions of down welling where cooler material sank but the whole surface was molten so these patterns were subtle variations in brightness and temperature rather than distinct solid features.

The young Earth had virtually no atmosphere during most of its formation, or at least no substantial atmosphere, because any gases present would have been blown away by impacts or by the intense solar wind from the young Sun, which was far more active and violent than the modern Sun, spewing out charged particles at rates perhaps a thousand times higher than today. Gases released from Earth's interior during impacts would briefly create temporary atmospheres of rock vapor and other volatiles, but these would quickly escape to space or condense back onto the surface as things cooled slightly between major impacts.

The composition of the material building Earth was not uniform, and this mattered for the planet's structure and future. The inner solar system contained two main types of solid material silicate rocks and metallic iron-nickel and they were not evenly mixed. Some planetesimals and embryos were more metal-rich, perhaps fragments of bodies that had already differentiated and then been shattered by collisions, exposing their iron cores. Others were more rock-rich, having formed from material that was naturally higher in silicates. As these bodies collided with the growing Earth, they delivered varying proportions of metal and rock, and the composition of Earth at any given moment depended on what happened to have hit it most recently.

On average, the material forming Earth was roughly 30 percent iron by mass and 70 percent silicate rock, though these are approximate figures and the exact ratio varied over time. This composition meant that once Earth grew large enough and hot enough for its interior to melt completely, the iron would separate from the rock and sink toward the center to form a core, while the lighter silicates would float upward to form the mantle. But this separation, this differentiation of Earth into layers, didn't happen all at once near the end of formation. It happened continuously throughout the accretion process, every time a giant impact delivered enough energy to melt the planet's interior completely.

Imagine what happened during a typical giant impact in the later stages of Earth's formation, when Earth had already accumulated perhaps 80 or 90 percent of its final mass and was struck by an embryo perhaps one-tenth Earth's mass a Mars-sized object traveling at ten or fifteen miles per second. The two bodies would have approached each other over hours or days, accelerating as they drew closer, until they finally collided with a relative velocity of several miles per second, releasing kinetic energy equivalent to perhaps a billion times the largest nuclear bombs humans have ever tested.

The impact point would have instantly vaporized, creating a fireball of rock vapor that expanded outward at tremendous speed, perhaps several miles per second, carving a crater thousands of miles across and hundreds of miles deep in Earth's surface though "surface" is a misleading term for a largely molten body. The shock wave from the impact would have propagated through Earth's interior at the speed of sound in rock, perhaps three or four miles per second, compressing material as it passed and heating it through compression, the same way a diesel engine heats air through compression until it's hot enough to ignite fuel.

Within minutes or hours, the shock wave would have circled Earth and converged at the point opposite the impact, the antipodes, creating a secondary zone of intense heating and disruption. The entire planet would have rung like a bell, vibrating at its fundamental frequencies for days or weeks after the impact. The collision would have added tremendous amounts of heat to Earth's interior, not just from the kinetic energy of the impact but from the subsequent mixing of hot material from the imp actor with Earth's interior, and from the compression caused by the shock wave.

The heat from this single impact would have been sufficient to melt Earth's interior completely, if it wasn't already molten, creating a global magma ocean extending from the surface down hundreds or thousands of miles, possibly all the way to the center depending on the violence of the impact and how hot Earth already was before the collision. And in this molten state, iron would separate from rock.

Iron is denser than silicate rock about twice as dense so when both are liquid and free to move, iron sinks and rock floats, driven by the same buoyancy forces that make oil float on water or air bubbles rise in a liquid. In Earth's magma ocean, droplets of molten iron would form and begin sinking toward the center, falling through the less dense molten silicates like rain falling through air, except moving much more slowly because molten rock is viscous and resists flow. The iron droplets might sink at speeds of perhaps a few feet per hour or a few feet per day, depending on their size and the viscosity of the surrounding magma, which means it would take thousands of years for iron to fall from the surface to the center, a journey of a few thousand miles through molten rock.

But as the iron sank, it released gravitational potential energy, and that energy was converted into heat through friction with the surrounding rock and through compression as the iron accumulated in the center. This process, called the iron catastrophe though it's not really a catastrophe but simply a natural consequence of differentiation, generated enormous amounts of additional heat, perhaps raising Earth's internal temperature by more than a thousand degrees. The sinking iron was heating Earth from within, adding to the heat from impacts and from the radioactive decay of elements like uranium and thorium that were mixed throughout the planet's interior.

The iron that reached the center accumulated into a growing core, a sphere of molten metal thousands of miles in diameter, denser and hotter than the surrounding rock. This core was not a single uniform mass but probably had complex internal structure and circulation patterns, with convection currents moving molten iron in complicated flows driven by heat escaping from the core's center toward its outer edge. The motion of this electrically conductive molten iron would later, once Earth had cooled and solidified enough for the flows to become organized, generate Earth's magnetic field through a dynamo process similar to how the moving wire in an electrical generator creates current. But during the formation period, when everything was chaotic and violently stirred by repeated impacts, there was probably no stable magnetic field, just transient magnetic disturbances that appeared and vanished as the interior was remixed by each new giant impact.

Above the iron core, the silicate rock formed Earth's mantle, a thick shell of molten or partially molten rock extending from the top of the core to the surface. The mantle was not composition ally uniform different types of silicate minerals have different densities and different melting points, so some degree of separation and layering would have occurred even within the mantle itself. But broadly speaking, the mantle was made of magnesium-iron silicates, compounds containing silicon, oxygen, magnesium, and iron in various combinations, along with smaller amounts of aluminum, calcium, and other elements.

The surface of this molten planet the boundary between the magma ocean and space was where heat was lost to the universe through radiation. Hot material at the surface radiated infrared light into space, cooling slightly in the process, and the cooled material would then sink back into the interior to be replaced by fresh hot material rising from below. This convective circulation was constantly turning over the magma ocean, mixing it and preventing any stable crust from forming for more than brief periods between impacts.

Occasionally, during a relatively quiet period between major impacts, the surface might cool enough for a thin solid crust to form, perhaps just a few miles thick, a fragile shell of solid rock floating on the magma ocean beneath. This crust would have been dark in color, probably black or dark gray, made of basaltic rock similar to the dark volcanic rocks we see on Earth today in places like Hawaii or Iceland. It would have been heavily fractured, broken into countless plates and chunks that shifted and collided with each other as the convicting magma beneath them pushed and pulled in different directions.

But these early crusts never lasted long. The next major impact would shatter them completely, mixing the solid rock back into the magma ocean, resetting the cooling process and returning Earth to a completely molten state. Earth probably formed and destroyed its first solid crust dozens or hundreds of times during the accretion period, each time cooling enough to solidify a thin surface layer, each time being remelted by the next giant impact. The planet was stuck in a cycle of partial solidification and complete remelting that would continue until the rate of major impacts decreased to the point where cooling could outpace heating.

The accretion process was not uniform in time. Computer simulations of planet formation suggest that the giant impact phase probably featured a few particularly large collisions separated by millions of years of relative calm punctuated by many smaller impacts. Early on, when there were many embryos in the inner solar system, collisions would have been relatively frequent, perhaps one major impact every few hundred thousand years. But as Earth grew larger and swept up most of the material in its orbital zone, the intervals between giant impacts would have lengthened because there were simply fewer large objects left to collide with.

The final stages of Earth's formation, the last ten million years or so before accretion essentially stopped, were probably characterized by increasingly long quiet periods interrupted by occasional catastrophic impacts as one of the last remaining planetary embryos finally crossed Earth's orbit and collided. These late impacts delivered the final five or ten percent of Earth's mass, adding material that would have mixed thoroughly with the existing planet during the global melting caused by each collision.

Throughout this entire violent formation period, Earth was accumulating not just rock and metal but also small amounts of volatile materials substances like water, carbon dioxide, nitrogen, and various organic compounds that have low boiling points and tend to exist as gases or liquids under most conditions. These volatile were present in the impacting planetesimals and embryos, incorporated into minerals or trapped in tiny pockets, and when delivered to Earth through impacts, they were released into the atmosphere or absorbed into the magma ocean or escaped to space depending on conditions at the time.

The exact amount of volatiles delivered during formation, and how much was retained versus lost to space, remains one of the major uncertainties in understanding Earth's formation. Water in particular is crucial for Earth's eventual habitability, yet we're not entirely certain how much was delivered during the primary accretion phase versus added later through impacts of water-rich objects from the outer solar system. The current leading theory suggests that most of Earth's water was present in the material that formed the planet the planetesimals and embryos in Earth's region of the solar system contained a few tenths of a percent water by mass bound in minerals or as ice inclusions, which is enough to account for Earth's oceans if most of it was retained during formation.

6

Fire, Magma, and a Shattered Surface

By the time Earth had accumulated most of its final mass perhaps ninety-five percent or more it had settled into a state that would persist for millions of years, a condition so extreme and alien that calling it hellish hardly captures the reality. The planet was entirely molten from its surface down through thousands of miles of depth, possibly all the way to its center, though the pressure at the deepest levels was so enormous that even material hot enough to melt at the surface remained solid in the core simply because the atoms were crushed too tightly together to flow freely. Earth was a ball of liquid rock roughly 8,000 miles in diameter, glowing dull orange or red from its own heat, rotating once every five or six hours much faster than today's twenty-four-hour rotation and surrounded by the faintest wisp of an atmosphere that was constantly being replenished by gases bubbling out of the magma ocean and constantly being stripped away by impacts and solar wind.

This was the magma ocean phase of Earth's history, and it's one of the most important and least understood periods in our planet's development because it set the stage for everything that would follow. The magma ocean determined how Earth's chemical elements were distributed between core, mantle, and surface. It controlled how volatiles like water and carbon dioxide were partitioned between the interior and any atmosphere that could form.

It established the initial conditions for the solid Earth that would eventually emerge once the magma ocean cooled and solidified. And yet we have no direct samples from this time, no rocks that survived from the magma ocean itself, because everything that existed then was later melted and re-melted and mixed and recycled through billions of years of geological activity.

What we know about the magma ocean comes from indirect evidence from computer simulations of planetary formation and cooling, from high-pressure experiments that recreate the conditions deep inside Earth, from the chemistry of the oldest rocks we can find which preserve hints about the composition of the mantle that produced them, and from theoretical physics and chemistry that predict how materials should behave under extreme conditions of temperature and pressure. The picture these different sources paint is consistent but not complete, filled with uncertainties and debates about details while the broad outlines seem reasonably solid.

The magma ocean was not a static pool of molten rock but a vigorously convicting system with powerful circulation patterns driven by the need to transport heat from the interior to the surface where it could be radiated away into space. Deep in the magma ocean, perhaps hundreds or thousands of miles below the surface, the rock was hotter heated by the decay of radioactive elements, by the residual heat from accretion impacts, and by the ongoing differentiation as iron continued to separate and sink toward the core. This hot rock was less dense than the cooler rock above it, creating buoyancy, and it began to rise, flowing upward in enormous currents that might have been hundreds of miles across.

As the hot rock rose toward the surface, it cooled slightly through expansion and through mixing with slightly cooler material around it, but it remained molten because the surface temperature of the magma ocean was still far above the melting point of rock, probably around 1,500 degrees Celsius or higher depending on the exact composition and the atmospheric pressure. When the rising current reached the surface, it would spread outward horizontally, pushed aside by more hot material rising from below, creating a diverging flow pattern similar to what happens when you pour honey onto a plate the honey spreads outward from the point of impact, pushed by the continuing stream from above.

At the surface, the hot rock radiated heat directly into space through thermal radiation, glowing in the infrared and visible spectrum, and as it cooled it became denser, eventually dense enough to sink back down into the interior. The cooled material would descend in other locations, perhaps along the boundaries between different rising currents, creating regions where surface material flowed downward back into the depths. These descending currents would be narrower and faster than the rising ones, concentrated into down welling zones where the surface appeared to be pulled downward, perhaps creating temporary depressions or trenches in the otherwise relatively smooth molten surface.

The timescale for this convection how long it took for rock to rise from the deep interior to the surface and then sink back down depended on the viscosity of the molten rock and the strength of the temperature differences driving the circulation. Molten rock is not like water; it's tremendously viscous, perhaps a million times thicker than water even when fully liquid, because the complex silicate molecules that make up rock don't flow easily past each other. This high viscosity meant that convection in the magma ocean was slow by everyday standards, with rising and sinking currents moving at speeds of perhaps a few feet per year or at most a few inches per day, taking tens of thousands of years or longer for rock to circulate from the deep interior to the surface and back.

But slow as it was, this convection was efficient at transporting heat because of the enormous temperature differences involved and because the vertical distances were so large. Heat was being continuously transported from Earth's interior, where temperatures might have exceeded 4,000 degrees Celsius in the deepest parts of the mantle, to the surface at 1,500 degrees or so, and then radiated away into the near-absolute-zero cold of space. Earth was cooling, gradually but inexorably, losing heat faster than it was generating new heat from radioactive decay and differentiation.

The cooling process was not smooth or uniform because of the continuing impacts. Every major impact, every collision with a planetesimal or embryo, added heat to the system both kinetic energy from the collision and the potential energy released as the impactor's material sank through the magma ocean to find its equilibrium level. Large impacts could temporarily reverse the cooling trend, adding so much heat that the magma ocean deepened or temperatures increased for a while before resuming their long-term decline. The planet was trying to cool down, but the universe kept heating it back up through impacts, creating a chaotic thermal history where the temperature rose and fell irregularly over millions of years, with the general trend downward but lots of variation and backtracking along the way.

As Earth cooled, different minerals began to crystallize from the magma ocean, solidifying wherever the temperature dropped below their melting points. The first minerals to crystallize were the ones with the highest melting points compounds rich in magnesium and iron, silicates with simple crystal structures that could form at temperatures of 1,900 degrees Celsius or higher. These early crystals were denser than the surrounding liquid magma, so they sank, falling through the convicting magma ocean like snow falling through air, accumulating at the bottom of the magma ocean at the boundary with the iron core.

This process of crystal formation and sinking is called fractional crystallization, and it's one of the primary ways that magma separates into different compositions, with different minerals forming at different temperatures and then physically separating from the remaining liquid. As the high-temperature minerals crystallized and sank, they removed certain elements from the magma ocean particularly magnesium and iron leaving the remaining liquid relatively enriched in other elements like silicon, aluminum, and calcium. This changed the composition of the magma ocean over time, making it evolve toward different mineral assemblages as cooling continued.

The crystals accumulating at the bottom of the magma ocean formed a growing layer of solid rock at the boundary between the liquid mantle and the solid iron core, creating what geologists call the cumulate layer because it accumulated from settling crystals. This layer was probably hundreds of miles thick, made of ultra-dense minerals that had sunk through the entire depth of the magma ocean, and it represented the first solid rock in Earth's mantle, the first permanent structure in the interior rather than just temporary features that were quickly remixed by convection or destroyed by impacts.

But the surface of the magma ocean remained liquid for a much longer time because the surface was being constantly refreshed by convection bringing hot material up from the interior, and because the surface was where impacts continued to deposit their energy. The surface couldn't solidify permanently until the interior had cooled enough that convection slowed down and the temperature difference between the interior and surface decreased to the point where surface material could cool below its solidification point before being swept away by convective currents.

This finally began to happen perhaps ten or twenty million years after Earth reached its final mass, once the impact rate had dropped to low enough levels that major impacts became rare, occurring perhaps only every few million years instead of every few hundred thousand years. The surface began to cool faster than the interior could reheat it, and solid rock began to form at the surface of the magma ocean, initially as small rafts and plates of solidified material floating on the liquid below, like ice forming on a lake in early winter.

These first crustal fragments were made of basalt, the same dark volcanic rock that forms today when magma erupts on Earth's surface and cools relatively quickly. Basalt is what you get when typical mantle-composition magma solidifies it's rich in iron and magnesium, dark gray or black in color, with small crystal grains because it cooled too quickly for large crystals to grow.

The basaltic crust forming on the magma ocean surface was probably only a few miles thick at most, maybe just a few hundred feet in some places, a fragile shell that could easily break and be re submerged into the magma ocean by convective currents or by impacts.

The formation of this earliest crust was not a single event but a stuttering, repetitive process that occurred over millions of years. Solid rock would form at the surface where cooling was most effective, thicken gradually as more magma crystallized beneath it, and then break apart due to stresses from the convicting magma below or from the differential cooling rates across the surface. Large plates of solid crust would crack and tilt and sink back into the magma ocean, melting as they descended into hotter regions, while new crust formed in other locations where fresh magma reached the surface and cooled.

The surface of early Earth during this transitional period must have been extraordinary to imagine, a patchwork of solid black crust broken into countless irregular plates and fragments, floating on a glowing orange sea of molten rock visible between the cracks and in regions where no solid crust had yet formed. The solid plates would have been in constant motion, pushed around by the convicting magma beneath them, colliding with each other and breaking apart, occasionally being sub ducted forced back down into the magma ocean—when one plate was pushed beneath another. It would have looked something like a frozen lake during the spring thaw, with ice plates breaking apart and jostling against each other, except the "water" between the ice would have been glowing molten rock at 1,400 degrees Celsius, and the whole scene would have been lit by the eerie orange glow from the exposed magma and the dull red heat radiated by the crust itself, which was still far too hot to touch or even approach.

The crust was not just forming at the surface through cooling but was also being created through volcanic activity, though volcanic activity during the magma ocean phase looked very different from modern volcanoes. There were no distinct volcanic mountains or cones because there was no stable surface to build them on, and no solid crust thick enough to contain pressurized magma beneath it. Instead, volcanism meant regions where magma up welling from the interior broke through the thin crust more vigorously than elsewhere, creating fountains or floods of molten rock that spread across the surface, adding new material to the crust wherever they flowed and solidified.

These volcanic floods were continuous in the sense that they occurred somewhere on Earth's surface at almost all times, driven by the never-ending convection in the magma ocean that constantly brought fresh hot material to the surface. The entire process of crust formation during this period was volcanic in nature every bit of solid rock at the surface originated as molten magma that had risen from the interior and cooled when exposed to space. The distinction we make today between volcanic rock and other types of rock didn't exist yet because there was only one way for rock to form: through the cooling and crystallization of magma.

As the crust gradually became more extensive and thicker, covering larger fractions of the surface and persisting for longer periods before being destroyed, Earth's appearance began to change. The planet was transitioning from a completely molten ball of liquid rock to something that had at least a partially solid surface, even if that surface was unstable and constantly being reworked. The color of the planet would have darkened as more of the glowing orange magma became hidden beneath the black basaltic crust, though the crust itself still glowed dull red from its own heat, and the many cracks and gaps in the crust would have blazed orange where molten rock remained exposed.

The thickness of the crust was limited during this period by the competition between cooling at the surface, which tried to solidify more rock and thicken the crust, and heating from below, which melted the bottom of the crust and thinned it. The base of the crust was in contact with the still-molten magma ocean, and at that boundary, the temperature was equal to the melting point of basalt around 1,200 to 1,300 degrees Celsius depending on the exact composition and pressure. Above that boundary, in the solid crust, the temperature decreased upward toward the surface, creating a temperature gradient through the crust.

The rate at which heat could be conducted through solid rock determined how thick the crust could grow before the heat flux from below balanced the heat loss at the surface. This equilibrium thickness was probably somewhere between a few miles and perhaps twenty or thirty miles, depending on when during the cooling process we're talking about and what the heat flux from the interior was at that time. Early on, when the mantle was hotter and convection was more vigorous, the crust probably couldn't grow much thicker than a few miles before the bottom melted as fast as the top solidified. Later, as the interior cooled and heat flux decreased, the crust could thicken to tens of miles in some regions.

But this early crust was composition ally monotonous it was all basalt or very similar compositions, all derived directly from the solidification of mantle-composition magma. The diverse rock types we see on Earth today granites, sandstones, limestones, shales didn't exist yet because they require more complex geological processes to form. Granite, for example, forms when basaltic crust is remelted and reprocessed, creating a different composition of magma that's richer in silicon and poorer in iron and magnesium. Sandstone forms from the weathering and erosion of existing rocks, a process that requires water and wind and time. Limestone forms from the accumulation of calcium carbonate, often produced by living organisms, and there were no organisms yet. The early Earth had only basalt and magma, a simple two-phase system with no compositional diversity at the surface.

The atmosphere during this period remained thin and transient, barely qualifying as an atmosphere at all compared to what would develop later. Gases were being released from the magma ocean continuously as volatile elements dissolved in the magma reached the surface and escaped primarily water vapor, carbon dioxide, nitrogen, and smaller amounts of sulfur compounds, hydrogen, and other gases. But these gases weren't accumulating to form a substantial atmosphere because they were being lost to space as fast as they were released, or faster.

The young Sun during this period was far more active than it is today, producing a much stronger stellar wind a stream of charged particles flowing outward from the Sun's surface and emitting far more ultraviolet radiation and X-rays. This intense radiation and particle flux interacted with any atmosphere Earth tried to form, heating the upper atmospheric layers to extremely high temperatures, thousands of degrees, giving the atmospheric molecules enough energy to escape Earth's gravity entirely. The process, called atmospheric escape, was efficient enough during the early solar system that even a planet as massive as Earth couldn't hold onto a substantial atmosphere of light gases like hydrogen and water vapor.

Additionally, the continuing impacts, even at reduced rates, would periodically blast away whatever atmosphere had managed to accumulate. When a large planetesimal struck Earth at ten or twenty miles per second, the impact vaporized not just the impactor and a large volume of Earth's crust, but also any atmosphere in the vicinity of the impact. The shock wave from the impact would propagate through the atmosphere, accelerating gas to velocities exceeding escape velocity, literally blowing the atmosphere off into space. Particularly large impacts could strip away the entire atmosphere temporarily, leaving Earth with essentially no gas envelope until volcanic outgassing from the magma ocean could begin rebuilding it.

So the early Earth's atmosphere was probably a thin, constantly fluctuating layer, thicker between major impacts and thinner immediately after them, composed mainly of rock vapor and volcanic gases, too hot and too tenuous to provide any significant pressure at the surface. There was no weather in any meaningful sense, no clouds or wind or storms, because the atmosphere was too thin and too hot and too unstable. The surface pressure might have been just a fraction of modern Earth's atmospheric pressure, perhaps one hundredth or one thousandth the current value, barely enough to notice if you could somehow stand on the surface without instantly vaporizing.

Deep inside the forming planet, hidden beneath thousands of miles of molten rock and solid crust, the iron core was continuing to develop and organize itself. The core had grown to roughly its final size about 2,200 miles in radius, containing roughly thirty percent of Earth's total mass but it was not yet the layered structure we see today with a solid inner core and liquid outer core. The entire core was molten, kept liquid by the intense heat from the planet's formation and from the ongoing release of gravitational energy as the core settled and organized itself.

The molten iron in the core was not pure iron but a mix of iron and nickel about 85 percent iron and 10 percent nickel by mass along with small amounts of lighter elements like sulfur, oxygen, silicon, and possibly hydrogen dissolved in the iron. These light elements were important because pure iron at the pressures found in Earth's core is actually denser than the observed density of the core, which means something lighter than iron must be mixed in to explain the measurements. The exact identity and abundance of these light elements in the core remains debated, one of the persistent uncertainties in our understanding of Earth's internal composition.

The core was convicting vigorously, with powerful flows of molten iron circulating between the deep interior and the outer boundary of the core at the base of the mantle. These convective motions were driven by heat escaping from the core into the overlying mantle, and by chemical buoyancy as lighter elements were concentrated in some regions and denser material in others. The convection patterns were probably chaotic and time-variable during this early period, constantly being disrupted by the continued accretion of iron sinking through the mantle and by the heat pulses from major impacts at the surface.

This convective motion of electrically conductive molten iron should, in principle, generate a magnetic field through the dynamo process, where moving conductor creates electrical currents, and those currents create magnetic fields that reinforce the motion. But it's unclear whether Earth had a stable, strong magnetic field during the magma ocean phase because the conditions were so chaotic and because we have no direct measurements or samples from this time that could tell us. Computer simulations suggest that a dynamo might have operated intermittently, producing a weak and unstable magnetic field that varied in strength and orientation over timescales of thousands or millions of years, nothing like the relatively stable field that Earth would develop later once the interior structure became more organized.

7

The Night the Moon Was Born

Earth was approximately forty to fifty million years old, give or take ten million years—the uncertainty is substantial because we're relying on radioactive dating of lunar samples and meteorites to constrain the timing, and the interpretation of those dates involves assumptions about the early solar system that might not be perfectly accurate. The planet had accumulated most or all of its final mass by this point, settling into what should have been the final stages of cooling from the magma ocean phase. The bombardment of planetesimals was winding down as Earth swept its orbital zone clear of large debris. The magma ocean was solidifying, with a growing solid mantle beneath an increasingly stable basaltic crust that now covered perhaps most of the surface, broken and fractured but persistent. Earth was transforming from a molten ball into something that resembled a proper planet with distinct layers and a solid surface, however hot and inhospitable that surface remained.

And then something catastrophic happened, something so violent that it nearly destroyed the planet that had spent tens of millions of years forming, something that reset Earth's thermal evolution and reshaped its structure in ways that still influence the planet today. Another planet collided with Earth in what may have been the largest impact in the inner solar system's history, and from the debris of that collision, the Moon was born.

We don't know what to call the impactor because it didn't survive the collision as a distinct body, but scientists have given it a provisional name borrowed from Greek mythology: Theia, after the Titaness who was the mother of Selene, the Moon goddess. Theia was not a small object, not just another planetesimal adding a fraction of a percent to Earth's mass. It was a planetary embryo in its own right, perhaps Mars-sized or slightly larger, containing roughly ten percent of Earth's mass, maybe as much as fifteen percent. It had formed in the inner solar system through the same accretion processes that built Earth, sweeping up its own feeding zone of planetesimals and smaller embryos, differentiating into a metallic iron core and a rocky silicate mantle, becoming a proto-planet that might eventually have become a proper terrestrial planet if it had been left alone to continue growing.

But Theia's orbit was not stable over long timescales. Gravitational interactions with other bodies in the inner solar system primarily Earth and Venus, the two largest planets in the inner system gradually perturbed Theia's orbit, changing its shape and orientation, pulling it into paths that brought it closer and closer to Earth over millions of years. The perturbations were small on each close approach, just slight modifications to the orbital parameters, but they accumulated over time the way small nudges can eventually topple a balanced object. Theia's orbit became more and more elliptical, swinging from farther out in the solar system down toward Earth's orbital distance, crossing Earth's path at an ever-increasing angle.

Eventually, inevitably, the two planets were in the same place at the same time. They had probably passed relatively close to each other dozens or hundreds of times over the preceding millions of years, close enough for their gravity to interact strongly but not close enough for a collision, near misses that further perturbed both orbits and set up the geometry for the final encounter. But this time, the closest approach was too close the separation between the two bodies was less than the sum of their radii, which means their surfaces made contact, which means they collided.

The approach was not head-on but came at an angle, with Theia striking Earth in what's called a grazing or oblique impact, perhaps hitting at an angle of thirty or forty degrees from horizontal rather than coming straight down vertically. This oblique geometry turned out to be crucial for what happened next, because a perfectly head-on collision between objects this large would likely have merged them completely, mixing them thoroughly, while a too-grazing collision might have resulted in Theia bouncing off or being captured into orbit intact. The actual impact angle was somewhere in between these extremes, violent enough to disrupt both bodies catastrophically but grazing enough to throw enormous amounts of material into orbit rather than having everything merge or escape.

The impact velocity was determined by gravity as Theia fell toward Earth, Earth's gravity accelerated it, adding speed beyond whatever velocity difference existed between their orbits. By the time of contact, Theia was probably traveling at something like ten miles per second relative to Earth's surface, possibly faster, perhaps as high as fifteen or twenty miles per second depending on the exact orbital geometry. This might not sound particularly fast compared to the speed of light, which is 186,000 miles per second, but at planetary scales with planetary masses, this velocity translated to kinetic energy beyond comprehension, perhaps equivalent to a trillion trillion tons of TNT, a number so large it becomes meaningless except as a mathematical abstraction.

When Theia's surface touched Earth's surface, the rocks at the contact point were instantly compressed to pressures far beyond anything that exists naturally in planetary interiors, millions of times atmospheric pressure, enough to change the atomic structure of the minerals and to heat them to tens of thousands of degrees through shock compression. The rock didn't just melt; it vaporized instantly, creating an expanding fireball of rock vapor at the impact site that grew to thousands of miles in diameter within seconds.

The shock wave from the impact a wall of compressed, super heated rock traveling faster than the speed of sound propagated through both bodies, traveling through Earth's interior at perhaps five or six miles per second and through Theia's interior at similar speeds. The shock wave carried tremendous energy, enough to melt or partially melt both planets' mantles completely as it passed, heating rock that might have been solid or only partially molten before the impact to temperatures well above their melting points. Within minutes, the shock wave had circled both planets, converging at the antipodes of the impact point on Earth's far side and creating a secondary zone of intense heating and fracturing.

Theia didn't just bounce off Earth or merge cleanly into it. Instead, the oblique angle of the collision meant that Theia was partially shattered and partially deflected, broken into large fragments by the impact while also having its trajectory bent by Earth's gravity. Parts of Theia perhaps the portion that hit Earth most directly merged with Earth, their material mixing thoroughly in the molten chaos at the impact site. But other parts of Theia, particularly material from the side that didn't directly contact Earth, were thrown outward at tremendous speeds by the impact, accelerated by the expansion of the vaporized rock at the collision point acting like an enormous explosion.

Similarly, a large volume of Earth's mantle at the impact site was excavated and thrown into space, accelerated by the same explosive expansion, creating a plume of molten and vaporized rock that expanded outward from Earth at speeds of several miles per second. This ejected material came primarily from Earth's outer layers the mantle and any crust that existed at the time because those were the parts that were hit and disrupted by the impact. Earth's core, deep in the interior, was largely protected from direct excavation, though the shock wave certainly affected it.

Within hours after the collision, the immediate violence had subsided enough to assess the damage, though of course there was no one present to do any assessing. Earth had survived as a distinct body, retaining most of its mass and its core, but its surface and outer mantle had been catastrophically disrupted. The entire planet was molten again, the magma ocean restored and deepened, possibly extending all the way to the core-mantle boundary, all the solidification and cooling of the previous tens of millions of years undone in a single afternoon. The surface glowed white-hot from the impact energy, radiating so much heat that it would have been far brighter than the Sun if viewed from space, though there was no one in space to view it.

Surrounding Earth was a vast cloud of debris fragments of both Theia and Earth, ranging in size from tiny droplets of molten rock to chunks perhaps hundreds of miles across, all orbiting Earth at various distances and velocities, some moving so fast they would escape Earth's gravity entirely and be lost to space, others moving slowly enough that they would fall back and re-impact Earth over the coming days and weeks. But a significant fraction of the debris perhaps fifteen to twenty percent of Earth's mass, though this number is quite uncertain had been thrown into orbits around Earth that were stable, at least temporarily, moving fast enough to avoid falling back immediately but not so fast as to escape.

The debris cloud surrounding Earth immediately after the giant impact was not the Moon, not yet, but rather the raw material from which the Moon would form over the coming months and years. This circumferential disk, as astronomers call it, extended from perhaps 10,000 miles above Earth's surface out to perhaps 30,000 or 40,000 miles, a vast doughnut of material orbiting the molten planet. The inner edge of the disk was limited by how close material could orbit without being torn apart by Earth's tidal forces the Roche limit, inside which the difference in gravitational pull between the near and far sides of an object exceeds the object's own gravity holding it together. The outer edge was less sharply defined, gradually thinning as distance increased, with the outermost material in barely bound orbits that might be perturbed by the Sun's gravity and lost over time.

The disk was not a uniform cloud but had considerable structure and variation. Some regions were denser than others, containing more material per unit volume, the result of how the debris was distributed during the initial ejection and how it subsequently clumped together through gravitational interactions. The debris was extremely hot, much of it molten or even vaporized, glowing brightly with its own heat and illuminated from below by the white-hot surface of Earth. If you could somehow have observed this scene from a safe distance in space though safe distances were hard to find so close to such an energetic event you would have seen Earth as a brilliant white sphere, possibly brighter than any star in the sky, surrounded by a glowing orange ring of debris that extended outward like the brim of a hat, the whole system rotating slowly, completing one revolution every five hours or so because Earth's rotation had been accelerated by the impact.

The debris in the disk was orbiting Earth, but these orbits were not stable or peaceful. Pieces of material were constantly colliding with each other as their orbits crossed, sometimes at low enough velocities that they stuck together, adding their mass and momentum, sometimes at high enough velocities that they shattered each other or splashed apart like liquid droplets colliding. The physics of these collisions depended on whether the material was solid, liquid, or vapor, and in the immediate aftermath of the impact, all three phases were present simultaneously in different parts of the disk.

The hottest, innermost regions of the disk were probably completely vaporized, existing as a gas of rock molecules and atoms, too hot for any condensed matter to exist. These rock vapors would have pressures and densities unlike anything we experience on Earth's surface, creating exotic conditions where silicates behave like atmospheric gases. As this vapor cooled through expansion and through radiating heat into space, it would eventually condense into liquid droplets, raining molten rock that fell inward toward the denser regions of the disk or coalesced into larger bodies through collisions.

The middle regions of the disk were probably a mix of molten droplets and solid fragments material that had cooled enough to solidify, or that had started solid and remained so because it wasn't heated sufficiently during ejection. These pieces ranged in size from dust particles to boulders to perhaps mountain-sized chunks in the largest cases, all orbiting Earth in slightly different paths, all being stirred and mixed by gravitational interactions and collisions, slowly settling toward the disk's mid plane where most of the mass would eventually concentrate.

The outer regions of the disk, farthest from Earth's heat and from the impact energy, were cooler and consisted primarily of solid fragments that had been thrown outward with enough velocity to reach these distant orbits. This outer disk would have been less dense than the inner regions, containing less mass spread over a larger volume, but it was still substantial, still containing enough material to build a significant moon if it could be gathered together.

The process by which this disk of debris transformed into the Moon is called accretion, the same fundamental process that built Earth and all the other planets, but operating on a much faster timescale because the debris was confined to a relatively small volume around Earth and because the individual pieces were orbiting in roughly the same plane, making collisions frequent. Computer simulations of moon formation suggest that the debris disk probably assembled itself into a single large moon plus perhaps a few smaller moonlets in somewhere between a few months and a few centuries after the impact, with most of the mass accumulating into the Moon within the first few years.

The assembly happened fastest in the densest parts of the disk, probably somewhere around 15,000 to 20,000 miles from Earth's center, where there was enough material for gravitational attraction to become significant between the larger fragments. A few of the largest pieces in this region perhaps originally hundreds of miles across, leftover chunks from Theia's mantle or large fragments of Earth's mantle that had been excavated by the impact would have served as seeds around which more material accumulated. These proto-moonlets swept up smaller debris through their gravitational pull, growing larger and extending their gravitational reach farther, pulling in more material in a runaway process.

As these proto-moonlets grew, they also began to interact with each other gravitation ally. Their orbits weren't perfectly circular or perfectly aligned, so they would periodically pass close to each other, exchanging momentum, sometimes colliding if they happened to be in exactly the wrong place at the wrong time. When two proto-moonlets collided at the relatively low velocities typical of objects in similar orbits around Earth perhaps a few hundred yards per second they would generally merge rather than shatter, because the collision energy was low enough that their gravity could hold the combined object together. Each merger created a larger body with stronger gravity, capable of sweeping up material from a larger volume of space.

Over time, through hundreds or thousands of such collisions and mergers, the largest body in the disk became dominant, accumulating perhaps 80 or 90 percent of the total mass in the debris cloud while a few smaller bodies contained the remainder. This largest body was the Moon, still growing but already recognizable as a distinct object, a spherical body perhaps 2,000 miles in diameter at this stage and continuing to grow as it swept up the remaining debris.

The Moon that emerged from this process was not the cool, gray, cratered object we see in the sky today. It was hot, intensely hot, glowing orange or even yellow from its own internal heat, possibly covered entirely by a magma ocean similar to what Earth had experienced during its own formation. The heat came from several sources the kinetic energy of all the collisions that had assembled the Moon, converted into heat when debris impacted the growing surface; the gravitational energy released as material fell through the Moon's gravitational field and hit the surface; and possibly some contribution from short-lived radioactive elements that were still decaying in the early solar system.

The Moon's magma ocean would have been hundreds of miles deep, possibly extending all the way to the Moon's center depending on how hot the initial assembly process was. Like Earth's magma ocean, it was a vigorously convicting system, with hot material rising from the interior and cooler material sinking back down, transporting heat from the interior to the surface where it could be radiated into space. The surface of the molten Moon would have been a glowing, churning sea of liquid rock, perhaps with some solid crust forming and breaking apart in the cooler regions, much like the early Earth's surface during its magma ocean phase.

But the Moon's magma ocean cooled much faster than Earth's had, for the simple reason that the Moon is much smaller. Heat is generated throughout a planet's volume in proportion to its mass but heat is lost through its surface area. For a spherical object, volume increases with the cube of the radius while surface area increases only with the square of the radius, which means smaller objects have higher surface-area-to-volume ratios and therefore cool faster. The Moon, with only about one-eightieth of Earth's mass and about one-quarter of Earth's radius, had roughly four times the surface area per unit mass compared to Earth, so it radiated its heat away about four times faster, cooling its magma ocean in perhaps a few million years rather than the tens of millions of years it took for Earth.

8

The First Atmosphere

After the Moon-forming impact and the subsequent cooling of both bodies from their molten states, Earth found itself in a position to develop something it had lacked throughout most of its violent formation a stable atmosphere that could persist for more than a few million years without being stripped away by impacts or blown off by the intense solar wind of the young Sun. The bombardment of large planetesimals had diminished to the point where major atmospheric-stripping impacts occurred only rarely, perhaps once every tens of millions of years rather than every few hundred thousand years. The Sun, while still more active than it is today, had settled into a more predictable pattern of behavior, its stellar wind weakening as the Sun matured. And Earth itself had grown massive enough that its gravity could hold onto an atmosphere of heavier gases even in the face of some atmospheric escape to space.

But the atmosphere that began to accumulate around Earth during this period, perhaps fifty to a hundred million years after the planet's formation, bore absolutely no resemblance to the air we breathe today. It was toxic, corrosive, and utterly incompatible with any form of life as we know it, though of course life didn't exist yet anywhere in the universe as far as we know, so this toxicity was irrelevant. There was no oxygen, or at most trace amounts produced by the splitting of water molecules through ultraviolet radiation from the Sun.

There was no ozone layer to shield the surface from that ultraviolet radiation. The atmosphere was thick with volcanic gases, dominated by compounds that we associate with volcanic eruptions and geothermal vents rather than with breathable air.

The source of this early atmosphere was Earth's interior, specifically the ongoing volcanic activity that brought molten rock from the mantle to the surface and released dissolved gases in the process. When rock melts deep underground, it can dissolve various volatile compounds substances that exist as gases at surface pressures within its liquid structure, much the way water can dissolve carbon dioxide to make carbonated beverages. The amount of volatiles that can dissolve in molten rock depends on the pressure and temperature, with higher pressures allowing more dissolution. Deep in Earth's mantle, where pressures are hundreds of thousands of times greater than atmospheric pressure, magma can hold significant quantities of dissolved water, carbon dioxide, sulfur compounds, and other gases.

When this volatile-rich magma rises toward the surface, either through volcanic eruptions or through the more gradual up welling of mantle material, the decreasing pressure allows the dissolved gases to come out of solution, forming bubbles within the magma, similar to how bubbles form in a carbonated drink when you release the pressure by opening the bottle. As the magma reaches the surface and erupts, these gases are released into the atmosphere, adding their mass to whatever atmospheric gases already existed.

This process of volcanic out gassing was the primary mechanism by which Earth's early atmosphere was created and maintained. Every volcanic eruption, every lava flow, every fumarole and hot spring released gases from Earth's interior into the atmosphere, slowly building up atmospheric pressure and changing atmospheric composition over millions of years. The rate of out gassing was probably much higher in the early Earth than it is today because the planet was hotter, which meant more vigorous convection in the mantle, which drove more volcanic activity at the surface. The early Earth may have had volcanic activity levels ten or a hundred times greater than modern Earth, with vast regions of the surface actively erupting or covered by fresh lava flows at any given time.

The composition of gases released by this volcanic activity can be inferred from several sources from the chemistry of modern volcanic gases, which gives us a baseline for what mantle-derived volatiles look like; from computer models of how different gases behave in magma at various pressures and temperatures; from the chemistry of ancient rocks that might preserve information about atmospheric composition through the minerals they contain or the isotopic ratios of their elements; and from theoretical considerations about what gases would have been present in the early solar system and how they would have been incorporated into Earth during its formation.

The dominant gas released by volcanic out gassing was almost certainly water vapor, H_2O, because oxygen is the third most abundant element in the universe and it preferentially combines with hydrogen to form water, making water by far the most abundant oxygen-bearing compound available. The early mantle probably contained somewhere between a few tenths of a percent and perhaps one percent water by mass, dissolved in minerals or present as trace hydroxyl groups bonded to silicate structures. This might not sound like much, but it represented an enormous reservoir one percent of the mantle's mass is enough water to fill Earth's current oceans several times over.

When mantle rock melted and rose to the surface, much of this water came with it, dissolved in the magma, and was then released as water vapor when the magma erupted. The early atmosphere therefore would have been rich in water vapor, perhaps containing more water vapor than any other gas, making it a thick, steamy envelope surrounding the planet. But this water vapor couldn't condense into liquid water and rain out of the atmosphere as long as the surface temperature remained above water's boiling point, which it probably did for some time after the giant impact due to the intense heat from that event and from the ongoing volcanic activity.

The second most abundant gas released by volcanic out gassing was probably carbon dioxide, CO_2, another oxygen-bearing compound that forms readily from the carbon and oxygen present in planetary materials. Carbon is less abundant than oxygen in cosmic terms, but it's still one of the more common heavy elements, and much of it ends up as carbon dioxide when released from the interior. Modern volcanic emissions are typically about five to ten percent carbon dioxide by volume, and the early Earth's volcanic gases probably had similar or higher proportions.

Carbon dioxide is important for planetary climate because it's a greenhouse gas, meaning it absorbs infrared radiation and traps heat in the atmosphere rather than allowing it to escape directly to space. A thick carbon dioxide atmosphere can keep a planet's surface much warmer than it would be with no atmosphere, which is why Venus, with its massive CO_2 atmosphere, has surface temperatures hot enough to melt lead despite being only slightly closer to the Sun than Earth. The early Earth probably had substantial amounts of carbon dioxide in its atmosphere, perhaps several bars of pressure worth several times the total pressure of Earth's current atmosphere creating a strong greenhouse effect that helped keep the surface warm.

Nitrogen, N_2, was probably the third major component of the early atmosphere, though unlike water and carbon dioxide, nitrogen doesn't come directly from volcanic out gassing in large quantities. Instead, nitrogen was probably delivered to Earth's atmosphere through a more complex pathway. During Earth's formation, nitrogen would have been incorporated into the planet mainly as nitrides compounds of nitrogen with metals or as ammonia, NH_3, bound in minerals or trapped in the interior. These nitrogen-bearing compounds are not very volatile and don't readily degas during normal volcanic activity. However, they can be broken down by high temperatures or by chemical reactions with other compounds, releasing nitrogen gas that can then escape to the atmosphere.

The most likely source of atmospheric nitrogen was probably ammonia photolysis ammonia being released by volcanic activity or by impacts and then broken apart by ultraviolet radiation from the Sun, which splits ammonia into nitrogen and hydrogen. The hydrogen, being very light, would have escaped to space relatively quickly, while the nitrogen, being heavier, would have accumulated in the atmosphere. Over time, this process could have built up substantial quantities of nitrogen gas, eventually making it one of the major atmospheric components.

Nitrogen is chemically inert under most conditions, meaning it doesn't react easily with other substances, so once it accumulated in the atmosphere it tended to stay there rather than being removed by chemical reactions with rocks or dissolved in oceans. This chemical stability is why nitrogen is the dominant gas in Earth's modern atmosphere it's not that nitrogen is particularly abundant or that it's being actively produced in large quantities, but rather that it accumulates over time because it doesn't get removed efficiently.

Beyond these three major components water vapor, carbon dioxide, and nitrogen the early atmosphere contained a variety of minor constituents that were nonetheless important for atmospheric chemistry and for the planet's subsequent evolution. Sulfur compounds were almost certainly present in significant quantities, released by volcanic activity in forms such as sulfur dioxide, SO_2, and hydrogen sulfide, H_2S, both of which are common in modern volcanic emissions. Sulfur dioxide reacts with water vapor in the atmosphere to form sulfuric acid, creating acidic droplets that would have made any rain that fell highly corrosive, capable of dissolving rocks and minerals much more effectively than pure water.

The early atmosphere also probably contained some methane, CH_4, and perhaps other simple hydrocarbons, though in smaller quantities than the major gases. Methane could have been produced by volcanic activity bringing reduced carbon compounds from the deep mantle, or by reactions between carbon dioxide and hydrogen in the presence of certain minerals. Methane is another greenhouse gas, more potent per molecule than carbon dioxide though typically present in smaller concentrations, and its presence would have contributed to keeping the surface warm. However, methane is unstable in the presence of oxygen and ultraviolet radiation, breaking down relatively quickly into carbon dioxide and water, so maintaining significant methane concentrations would have required continuous production to replace what was being destroyed.

Hydrogen, H_2, was probably present as well, perhaps in small but measurable quantities, released both by volcanic out gassing and by the breakdown of water molecules through photo dissociation ultraviolet radiation from the Sun splitting water into hydrogen and oxygen. But hydrogen is the lightest element and escapes Earth's gravity relatively easily, so it wouldn't have accumulated to high concentrations. Any hydrogen released into the atmosphere would have gradually worked its way to the upper atmosphere where it could achieve escape velocity and be lost to space, carrying away small amounts of mass continuously over millions of years.

The absence of free oxygen in the early atmosphere is worth emphasizing because it's such a departure from modern Earth and because it had profound consequences for the chemistry of the surface and the potential for life. Oxygen is highly reactive it readily combines with almost everything it touches, oxidizing metals, breaking down organic compounds, participating in combustion. The modern Earth's atmosphere is about 21 percent oxygen by volume, and this oxygen is constantly reacting with rocks, minerals, organic matter, and anything else exposed to it. The only reason oxygen remains abundant in the modern atmosphere is that it's being continuously replenished by photosynthetic organisms plants, algae, and photosynthetic bacteria that split water molecules and release oxygen as a waste product.

Without life to produce it, free oxygen doesn't accumulate in planetary atmospheres because it gets removed too quickly by reactions with surface materials. Any oxygen that formed in the early atmosphere through the photo dissociation of water vapor ultraviolet light splitting H_2O into hydrogen and oxygen would have quickly reacted with volcanic gases like hydrogen sulfide or methane, or with reduced minerals at the surface like iron-bearing rocks, forming oxides and being removed from the atmosphere. The early Earth was what chemists call a reducing environment, dominated by gases and minerals that wanted to donate electrons rather than accept them, the opposite of an oxidizing environment like modern Earth's surface.

This reducing atmosphere had several important implications. First, it meant that any organic molecules that formed carbon-based compounds that are the building blocks of life would have been more stable than they are in the modern oxygen-rich atmosphere, because oxygen would have broken them down through oxidation. Second, it meant that the minerals forming at the surface would have been different from modern minerals, with iron existing primarily in its reduced form, Fe^{2+}, rather than the oxidized form, Fe^{3+}, that dominates modern sediments and soils. Third, it meant that any liquid water that existed would have been chemically very different from modern oceans, capable of dissolving minerals and metals that are insoluble in oxygenated water.

The pressure and density of this early atmosphere are difficult to constrain precisely, but we can make reasonable estimates based on how much gas could have been released from the interior over the first few hundred million years of Earth's history and how much would have been lost to space through various escape mechanisms. The total pressure at the surface was probably somewhere between a fraction of modern atmospheric pressure and several times modern pressure, with most estimates clustering around one to three times the current sea-level pressure of about one bar, or 14.7 pounds per square inch.

A denser atmosphere than modern Earth would have had several effects on the planet's conditions. It would have made the greenhouse effect stronger, trapping more heat and keeping the surface warmer even though the young

Sun was dimmer than the modern Sun a crucial point we'll return to shortly. It would have made the atmosphere more opaque, blocking more sunlight from reaching the surface but also blocking more infrared radiation from escaping to space. And it would have created higher surface pressures that affected the behavior of water raising the boiling point of water and allowing liquid water to exist at higher temperatures than it could under lower pressure.

The color and appearance of this early atmosphere would have been utterly unlike the clear blue sky we see today. The thick water vapor would have made the atmosphere hazy and opaque, possibly creating a whitish or grayish appearance from space as water droplets scattered sunlight in all directions. The sulfur compounds would have contributed yellowish or brownish tints, particularly in regions downwind of active volcanoes where sulfur dioxide concentrations were highest. The overall effect would have been a murky, turbulent atmosphere, possibly with multiple layers of clouds at different altitudes, all of it shrouding the surface in a thick blanket of gases that blocked most direct sunlight from reaching the ground.

Lightning would have been common in this early atmosphere, perhaps even more common than on modern Earth, because the vigorous convection driven by the volcanic heat and the greenhouse warming would have created powerful storms with strong updrafts and downdrafts. Lightning in an atmosphere containing methane, ammonia, water vapor, and various other compounds can drive interesting chemistry, creating complex organic molecules from simpler precursors through the energy of electrical discharges. Scientists have demonstrated in laboratory experiments most famously the Miller-Urey experiment in 1953 that passing electrical sparks through mixtures of gases similar to what might have existed in the early atmosphere can produce amino acids and other organic compounds, the building blocks of proteins and other biological molecules.

One of the most significant challenges for understanding Earth's early atmosphere is explaining how the surface could have been warm enough for liquid water to exist despite the faint young Sun problem. The Sun, like all stars, gradually brightens as it ages because the helium building up in its core from hydrogen fusion changes the core's structure and temperature in ways that increase the rate of fusion. Computer models of stellar evolution indicate that the Sun was probably only about 70 to 75 percent as luminous four billion years ago as it is today, delivering significantly less energy to Earth. If Earth had its current atmosphere composition and received 25 to 30 percent less sunlight, the planet would have been frozen, with average surface temperatures well below the freezing point of water.

Yet we have strong evidence that liquid water existed on Earth's surface very early in its history, probably by four billion years ago and possibly earlier, which means something must have kept the surface warm despite the dimmer Sun. The most likely explanation is a much stronger greenhouse effect than exists today, primarily from high concentrations of carbon dioxide and water vapor, possibly supplemented by methane and other greenhouse gases. An atmosphere containing a hundred or a thousand times more carbon dioxide than the modern atmosphere could have trapped enough heat to keep the surface above freezing even with the reduced solar input, solving the faint young Sun problem through atmospheric composition rather than requiring any exotic explanations.

9

The Arrival of Water

Earth had accumulated most of its mass, survived the catastrophic Moon-forming impact, developed a stable if hostile atmosphere, and was cooling from its magma ocean state toward something that could maintain a solid crust for more than brief intervals between impacts. The surface temperature was dropping, no longer hot enough to melt rock everywhere, though volcanic activity remained intense and localized hot spots still glowed with molten lava. The atmosphere, thick with water vapor and carbon dioxide and various other volcanic gases, pressed down on the surface with crushing weight, creating pressures perhaps several times higher than modern atmospheric pressure. And then, somewhere around four billion years ago or perhaps slightly earlier, something remarkable happened that would fundamentally transform Earth and make it unique among all known planets—liquid water began to collect on the surface in quantities large enough to form the first oceans.

The story of how Earth acquired its water is more complex and uncertain than you might expect, because water doesn't leave a clear signature of its origins that we can easily read billions of years later. We know Earth has substantial water enough to cover the planet to a depth of about 1.7 miles if spread evenly across the surface, or about 0.02 percent of Earth's total mass but we're less certain about where all that water came from or when it arrived.

The water could have been present in the materials that formed Earth from the beginning, incorporated into planetesimals and embryos as they accredited into our planet. Or it could have been delivered later, after Earth formed, through impacts of water-rich objects from the outer solar system where ice was stable. Or, most likely, some combination of both sources contributed, with the relative importance of each remaining a subject of active research and scientific debate.

The planetesimals and embryos that collided to build Earth in the inner solar system formed in a region where temperatures were too high for water ice to condense directly from the solar nebula this was inside the frost line, that boundary beyond which water could exist as ice but inside which it could only exist as vapor or as tiny amounts bound in minerals. However, water wasn't entirely absent from this inner region. Some minerals can incorporate water or hydroxyl groups into their crystal structures even at temperatures above water's freezing point, effectively trapping water within rocky material that doesn't look wet in any obvious way. These hydrated minerals compounds like serpentine or amphibole could have been present in the planetesimals forming Earth, containing perhaps a few tenths of a percent water by weight.

Additionally, the boundary between the inner dry region and the outer icy region wasn't perfectly sharp or perfectly stable. The frost line moved inward and outward over time as the young Sun's luminosity varied, and material from beyond the frost line could be scattered inward through gravitational interactions with forming planets, delivering some ice-rich material to the inner solar system even though ice couldn't form there directly. Comets and icy asteroids from the outer asteroid belt or beyond could have been perturbed into orbits that brought them into the inner solar system where they eventually collided with Earth, delivering their water content despite having formed in much colder regions.

The current scientific consensus, based on multiple lines of evidence, suggests that most of Earth's water probably came from the materials that formed the planet rather than being added later through cometary impacts, though comets and asteroids may have contributed a smaller secondary component. The evidence for this conclusion comes primarily from measuring the isotopic composition of Earth's water specifically, the ratio of deuterium to normal hydrogen and comparing it to the isotopic ratios measured in various potential water sources like comets, asteroids, and meteorites.

Deuterium is a heavy isotope of hydrogen containing one proton and one neutron in its nucleus instead of just the single proton found in normal hydrogen. Water molecules can contain either normal hydrogen or deuterium, creating regular water, H_2O, or heavy water with one or two deuterium atoms replacing the normal hydrogen. The ratio of deuterium to hydrogen varies slightly between different regions of the solar system and different types of objects, depending on where and how the water formed and what processes it experienced. By measuring this ratio in Earth's oceans and comparing it to measurements from comets and meteorites, scientists can identify which potential sources match Earth's water most closely.

Most comets measured to date have deuterium-to-hydrogen ratios significantly higher than Earth's oceans, suggesting that cometary water can't be the dominant source of Earth's water, or at least that the comets we've measured aren't representative of the ones that might have hit early Earth. Certain types of meteorites, particularly a class called carbonaceous chondrites that come from the outer asteroid belt and contain significant amounts of water bound in minerals, have deuterium ratios much closer to Earth's values, suggesting these asteroid-like bodies were probably the main source of Earth's water, either delivered as intact projectiles after Earth formed or incorporated into the planet during its initial accretion.

Regardless of exactly how the water arrived or when, what matters for our story is that Earth had accumulated substantial amounts of water perhaps an ocean's worth or more locked in its interior during and shortly after its formation. This water was dissolved in the mantle, bound in minerals, or possibly existing as pockets of fluid in the deep interior where pressures kept it from escaping. The water wasn't at the surface yet because the surface was too hot; any water that reached the surface during the magma ocean phase would have immediately vaporized, becoming part of the atmosphere as steam, unable to condense into liquid as long as the surface temperature remained above water's boiling point.

But as Earth cooled following the Moon-forming impact and the subsequent gradual decline in volcanic activity and impact frequency, the surface temperature eventually dropped below a critical threshold. We can't pinpoint exactly when this happened the timing depends on too many uncertain factors like the greenhouse effect strength, the volcanic out gassing rate, the residual heat from formation, and the solar luminosity at the time. But somewhere between 4.4 and 4.0 billion years ago, probably closer to 4.2 or 4.1 billion years ago, the surface temperature fell below the boiling point of water at the pressures existing in the atmosphere.

This meant that water vapor in the atmosphere could finally condense into liquid droplets, the same way water vapor condenses into dew on grass when the temperature drops below the dew point. But this wasn't just a light morning dew forming on a planetary surface this was a massive phase transition involving an entire ocean's worth of water that had been suspended as vapor in the atmosphere, all of it ready to condense as soon as the temperature dropped sufficiently. The atmosphere had been holding this water in gaseous form for millions of years, unable to let it rain out because the surface was too hot, but now the conditions had finally changed enough that liquid water could exist at the surface.

And so it began to rain.

Not a brief shower or even a prolonged storm, but rainfall that may have continued for hundreds of thousands of years, possibly even millions of years, as the atmosphere gradually released its burden of water vapor. The rain fell on hot rock the surface was still far above room temperature even if it was below water's boiling point and much of it immediately boiled back into steam, re-entering the atmosphere only to cool and condense and fall again in an endless cycle. The surface would have been shrouded in fog and steam, a boundary layer between the hot rock and the cooler atmosphere above, creating conditions somewhere between a sauna and a pressure cooker, with water constantly transitioning between liquid and vapor.

The earliest rainfall on Earth was nothing like the gentle or even violent rainstorms we experience today. The rain fell through an atmosphere that was denser and hotter than modern air, filled with volcanic gases and thick with clouds, creating droplets that may have been larger than modern raindrops because of the higher atmospheric pressure allowing bigger drops to form before they fell. The rain was acidic, contaminated with sulfur dioxide and other volcanic gases that dissolved into the water droplets as they fell, creating sulfuric acid and other compounds that made the rain chemically aggressive, capable of dissolving rocks and minerals far more readily than pure water could.

When this hot, acidic rain struck the surface, it didn't simply pool in low spots and sit there peacefully. The surface was still above the boiling point in many locations, so the water flashed back into steam instantly, creating hissing clouds of vapor that rose back into the atmosphere. In slightly cooler areas, the water might have remained liquid for a few moments or a few minutes before boiling away, just long enough to flow downhill in tiny rivulets, beginning the process of erosion that would eventually reshape the planet's surface. In the coolest regions perhaps high altitude areas or regions far from active volcanic zones the water might have pooled briefly in small puddles or ponds before evaporating back into the atmosphere.

This cycle of rain, boiling, evaporation, condensation, and more rain continued for a very long time, gradually cooling the surface through the removal of heat. Every time water evaporated from the hot surface, it carried thermal energy with it, transporting that heat upward into the atmosphere where it could be radiated away to space more effectively than heat could be radiated directly from the surface. Water is extraordinarily efficient at moving heat because of its high heat capacity it takes a large amount of energy to heat water or to evaporate it and because phase changes between liquid and vapor involve large energy transfers. The water cycle was acting as a massive heat engine, pumping thermal energy from Earth's surface into the upper atmosphere and eventually into space.

As the surface continued to cool, more and more of the rain remained liquid for longer periods, flowing farther before evaporating, carving deeper channels, creating larger pools. The lowest areas on the surface wherever tectonic processes or impacts had created depressions began to accumulate water that didn't immediately boil away, forming the first persistent bodies of liquid water, though "persistent" at this stage might mean hours or days rather than years or centuries. These proto-lakes were shallow and temporary, constantly shrinking through evaporation and being replenished by new rainfall, their shorelines advancing and retreating with the weather patterns and with the local volcanic activity that continued to heat some regions while leaving others cooler.

The water flowing across the surface began to reshape the landscape in ways that had never occurred during the magma ocean phase. Liquid water is a powerful erosive agent, capable of dissolving minerals, loosening rock particles, and carrying sediment from high elevations to low elevations through the simple action of gravity pulling the water downhill. The flowing water carved channels and valleys, however crude and temporary, creating drainage patterns that concentrated water flow in certain paths and left other areas high and dry. The sediment carried by this water small particles of rock and minerals eroded from the highlands was deposited in the lowlands, beginning the process of creating layered sedimentary deposits, though these earliest sediments were probably quickly buried by volcanic flows or destroyed by impacts.

The chemical interactions between water and rock were equally important as the physical erosion. When water contacts silicate minerals the compounds that make up most of Earth's crust and mantle it can react with them through a process called hydrolysis, breaking chemical bonds in the minerals and forming new compounds. These reactions are slow at low temperatures but proceed much faster at the elevated temperatures that characterized early Earth's surface. The hot, acidic water flooding across the basaltic crust would have reacted vigorously with the minerals it contacted, dissolving some components while leaving others behind, altering the chemical composition of both the water and the rock.

One particularly important reaction was the weathering of silicate minerals to form clay minerals hydrated aluminum silicates that are stable at surface conditions and that form the basis of most modern soils. When basalt weathers in the presence of water, the calcium, sodium, magnesium, and iron in the rock tend to dissolve into the water, leaving behind aluminum and silicon combined with water molecules to form clays. This weathering process also consumes carbon dioxide from the atmosphere the carbonic acid formed when CO_2 dissolves in water helps break down the silicate minerals, and the dissolved cations like calcium and magnesium eventually combine with carbonate ions to form solid carbonate minerals like limestone.

This weathering reaction provided a crucial negative feedback mechanism for Earth's climate, a natural thermostat that helped regulate surface temperature over long timescales. When Earth was hot and wet, weathering reactions proceeded rapidly, pulling carbon dioxide out of the atmosphere and storing it in carbonate minerals, which weakened the greenhouse effect and allowed the planet to cool. When Earth cooled too much, weathering slowed because there was less liquid water or because the reactions proceeded more slowly at lower temperatures, allowing volcanic out gassing to build up carbon dioxide in the atmosphere and strengthen the greenhouse effect, warming the planet. This feedback system, sometimes called the carbonate-silicate cycle, has helped maintain Earth's surface temperature within the range where liquid water can exist for billions of years despite the Sun's gradually increasing luminosity.

As the surface continued cooling and as more water accumulated in liquid form rather than remaining as atmospheric vapor, the atmospheric composition began to change significantly. The water vapor content of the atmosphere decreased dramatically as water condensed and rained out, falling from perhaps a substantial fraction of the atmospheric mass to much smaller amounts. This reduced the atmospheric greenhouse effect because water vapor is a powerful greenhouse gas, which in turn allowed more cooling, which caused more water to condense out, creating a positive feedback loop sometimes called a runaway condensation that rapidly transferred water from the atmosphere to the surface once conditions crossed the threshold where condensation could begin.

The atmospheric pressure also decreased as water condensed out, because the water vapor that had been contributing to the total pressure was now liquid at the surface rather than gas in the atmosphere. The atmosphere became less opaque, less filled with clouds and water droplets, allowing more sunlight to penetrate to the surface and making the sky potentially visible for the first time since the planet formed. Though the sky wouldn't have been the familiar blue of modern Earth it was still filled with volcanic gases and possibly tinted by sulfur compounds and other contaminants at least there was beginning to be a sky rather than just an impenetrable blanket of clouds and steam.

The transition from a steam-filled atmosphere to one with liquid water at the surface probably didn't occur uniformly across the whole planet at once. Different regions cooled at different rates depending on their elevation, their distance from volcanic activity, and their latitude and relationship to atmospheric circulation patterns. High elevation areas would have cooled first because temperature decreases with altitude in an atmosphere, so mountain peaks if any existed at this early stage would have been the first places where rain fell and didn't immediately boil away. Polar regions would have been cooler than equatorial regions because they received less direct sunlight, so the first permanent bodies of water might have formed at high latitudes.

But eventually, over hundreds of thousands or millions of years of continuous rainfall and surface cooling, liquid water became widespread across Earth's surface, collecting in the lowest areas to form the first oceans. These earliest oceans were probably not very deep perhaps a few hundred feet or at most a few thousand feet depending on how much water had accumulated and how much of the surface was covered but they represented a fundamental transformation of Earth's surface conditions from a dry, hot world to a wet, cooler one.

10

A Cooling World

Earth had crossed a threshold, moved past a boundary that separated the violent, molten chaos of its formation from something more stable and structured, more recognizable as a planet in the modern sense. The surface was no longer universally molten, no longer everywhere glowing with its own heat, no longer being constantly remelted by giant impacts that delivered enough energy to maintain global magma oceans. Instead, solid rock now covered most or all of the surface, broken and fractured and still occasionally disrupted by volcanic activity, but solid nonetheless, forming a permanent crust that could persist for millions of years rather than being destroyed and reformed on shorter timescales. Liquid water was pooling in the low-lying areas, creating bodies of water that ranged from small ponds to vast proto-oceans covering significant fractions of the planet's surface. The atmosphere had thinned as water vapor condensed out, becoming less opaque, less dominated by steam, though still thick with carbon dioxide and volcanic gases that created a persistent haze.

The planet was cooling, steadily and irreversibly, shedding the heat accumulated during its formation and finding a new thermal equilibrium governed by the balance between heat sources and heat losses. The heat sources were fewer and less intense than they had been during the formation period radioactive decay of elements like uranium, thorium, and potassium in Earth's interior continued to generate warmth, contributing perhaps a few dozen terawatts of power, but this was far less than the energy that had been delivered by impacts during the heavy bombardment. The residual heat from Earth's formation, the thermal energy stored in a planet that had been heated to melting throughout much of its interior, was gradually working its way to the surface through convection and conduction, but this too was declining over time as the planet radiated energy away into the cold void of space.

The cooling process was not smooth or uniform but varied dramatically across different depths and regions within the planet. The surface cooled fastest because it was in direct contact with space, or at least with the atmosphere which was in contact with space, allowing heat to be radiated away as infrared light. The surface temperature dropped from perhaps 1,500 degrees Celsius during the late magma ocean phase to perhaps a few hundred degrees or even cooler in some regions by around four billion years ago, crossing below the boiling point of water and eventually settling toward temperatures where liquid water could exist comfortably without constantly threatening to boil away. The exact surface temperature at any given time and location would have varied enormously depending on volcanic activity, atmospheric greenhouse effects, solar heating, and latitude, but the general trend was unmistakably downward.

The upper mantle, just beneath the crust, cooled more slowly than the surface because it was insulated by the overlying rock and because it was being heated from below by the deeper mantle. But it too was gradually solidifying from a partially molten state toward being fully solid, or at least solid enough that only a small fraction of the rock remained molten in isolated pockets. The transition from a mostly molten upper mantle to a mostly solid one fundamentally changed how heat was transported from the interior to the surface, shifting from a regime where molten rock could convict relatively easily, flowing in large-scale circulation patterns, to one where heat had to move primarily through solid rock, either by conduction the slow transfer of heat from atom to atom through a solid material or by solid-state convection, where solid rock can flow very slowly over millions of years when subjected to sustained stress and temperature differences.

This transition to a solid or nearly solid mantle had profound consequences for Earth's volcanic activity and for the eventual development of plate tectonics. When the mantle was mostly molten, volcanism was relatively easy molten rock could rise to the surface through fractures or weak zones with minimal resistance, creating frequent and widespread volcanic eruptions. But as the mantle solidified, volcanism became more difficult and more localized, requiring either hot spots where the mantle remained particularly hot and partially molten, or regions where tectonic stresses created cracks and decompression that allowed solid rock to melt and rise. The character of volcanic activity began to shift from continuous, widespread lava floods covering vast areas to more discrete volcanic centers separated by regions with little or no volcanic activity.

The lower mantle and core, deep in Earth's interior where pressures were enormous and temperatures remained extremely high, cooled even more slowly than the upper mantle. The core, that sphere of molten iron-nickel alloy at Earth's center, was losing heat to the overlying mantle, but the rate of heat loss was limited by how quickly the mantle could transport heat away through convection and conduction. The temperature at the core-mantle boundary the interface between the solid silicate rock of the lowermost mantle and the liquid metal of the outer core might have been around 4,000 degrees Celsius or higher during this early period, hundreds of degrees hotter than it would be billions of years later.

As the core cooled, something important was happening to its structure. The very center of the core, where pressures reached millions of times atmospheric pressure, was beginning to solidify despite the intense heat. This might seem paradoxical how can the hottest part of Earth be solid? but it occurs because the melting temperature of iron increases with pressure faster than the actual temperature increases with depth. At the extreme pressures in Earth's center, iron's melting point is pushed above the actual temperature, causing iron to crystallize into a solid even though it would be molten at lower pressures.

This growing inner core, forming at Earth's center through the crystallization of iron from the liquid outer core, was probably very small during this early period, perhaps just a few hundred miles in radius or possibly not yet formed at all depending on exactly how hot the core was and how the temperature profile compared to the pressure-dependent melting curve of iron. But it was beginning to form or would form soon, growing slowly over billions of years as the core continued to cool, and its formation would eventually become important for Earth's magnetic field because the crystallization process releases heat and changes the composition of the remaining liquid iron, driving convection in the outer core that generates the magnetic field through dynamo action.

The magnetic field itself during this early period remains mysterious and contested among scientists. We don't have direct measurements of the ancient magnetic field strength or structure, obviously, and the indirect evidence from ancient rocks is limited and difficult to interpret. Some rocks dated to around 3.5 or 3.4 billion years ago show signs of having formed in the presence of a magnetic field of roughly modern strength, suggesting Earth had a functioning dynamo by that time. But whether a magnetic field existed earlier, during the period we're discussing around four billion years ago, remains uncertain. The conditions for generating a magnetic field a convicting, electrically conducting fluid in the core probably existed, but whether the convection was organized and persistent enough to create a stable dipole field or whether it produced only weak, chaotic magnetic fields is unclear.

The crust that formed during this cooling period was primarily basaltic in composition, similar to the oceanic crust that forms today at mid-ocean ridges where mantle rock rises and melts due to decompression, creating basalt that solidifies to form new seafloor. This basaltic crust was dark in color, relatively dense compared to other types of crust that would form later, and rich in iron and magnesium compared to the lighter, more silicon-rich rocks like granite that would eventually dominate the continents. The thickness of this early crust is uncertain but was probably somewhere between five and ten miles in most locations, perhaps thicker in some areas and thinner in others depending on local volcanic activity and crustal formation processes.

The early crust was not a single, continuous shell encasing the planet but rather a collection of plates or fragments that floated on the partially molten or very hot solid mantle beneath them, moving relative to each other in response to the convective motions in the mantle and to their own thermal contraction as they cooled. This movement was probably more vigorous and chaotic than modern plate tectonics, driven by the much higher heat flux from the hotter interior, creating a surface that was constantly being rearranged and recycled on timescales of millions of years or even faster.

Where two crustal fragments moved apart, pulled by the underlying mantle convection, fresh magma would well up from below to fill the gap, solidifying to create new crust in a process similar to modern seafloor spreading. Where fragments collided, being pushed together by mantle currents, one would typically be forced beneath the other in a crude form of subduction, sinking back into the hot mantle where it would partially or completely remelt, its material recycled back into the mantle to eventually rise again in some other location. This recycling meant that very little of Earth's earliest crust has survived to the present day it was all destroyed through subduction and remelted, making it extraordinarily difficult to study the conditions on Earth during the first few hundred million years of its history because the direct rock record from that time has been almost entirely erased.

The small amount of ancient crust that has survived, the oldest rocks we can find on Earth today, dates back to only about four billion years ago, and even these rocks have been heavily modified by heat, pressure, and chemical alteration over the billions of years since they formed. The very oldest mineral grains we can identify, tiny crystals of zircon found within younger rocks in Western Australia, date back to about 4.4 billion years ago, providing our earliest direct samples of Earth materials. These zircon crystals, each one smaller than a grain of sand, contain oxygen isotopes that suggest they crystallized from magma that had interacted with liquid water, providing some of our best evidence that oceans or at least substantial bodies of water existed by 4.4 billion years ago, remarkably early in Earth's history.

The presence of liquid water at the surface had profound effects on the cooling and evolution of the crust and mantle. Water is not just a passive substance sitting in oceans it's chemically active and mobile, capable of infiltrating into rocks through cracks and pore spaces, participating in chemical reactions, and being carried deep into the planet through subduction or other processes. When water-bearing rocks are carried down into the mantle through subduction, the increasing temperature and pressure eventually drive the water out of the minerals that were holding it, releasing the water into the surrounding mantle rock. This water, even in small amounts, dramatically lowers the melting point of mantle rock, causing it to melt at temperatures hundreds of degrees cooler than dry rock would melt.

This water-induced melting in the mantle created magmas with different compositions than would form through the melting of dry mantle rock. The magmas were richer in silicon and poorer in iron and magnesium, compositions that would eventually lead to the formation of continental crust the lighter, thicker, more silicon-rich crust that forms the continents as opposed to the denser basaltic oceanic crust. The process probably began quite early, perhaps as soon as there was enough water at the surface to be carried down into the mantle through subduction, but it took time to build up significant volumes of continental-type crust because the process was inefficient and because much of what formed was later destroyed and recycled.

The formation of this proto-continental crust, even in small amounts, created a new type of terrain on Earth's surface regions that were lighter and therefore more buoyant than the surrounding basaltic crust, standing higher above the mantle and therefore poking up above sea level to form the first land areas, the first places on Earth where rock was exposed to air rather than being submerged beneath water. These earliest landmasses were probably small, perhaps more like volcanic islands than proper continents, scattered across an otherwise water-covered globe, but they represented something fundamentally new in Earth's history the boundary between ocean and atmosphere, the shoreline, the place where water, rock, and air all met and interacted.

The existence of exposed land, even small amounts of it, changed the surface chemistry and weathering patterns. Rock exposed to air and rain weathers differently than rock submerged in seawater, experiencing more dramatic temperature fluctuations between day and night and between seasons, being subjected to physical weathering from thermal expansion and contraction, and reacting with atmospheric gases that dissolved in rainwater to create acidic solutions. The weathering products clay minerals, dissolved ions, sediment particles were washed off the land by rainfall and streams, carrying them into the oceans where they accumulated as sediments on the seafloor or dissolved into the seawater, gradually changing the chemical composition of the oceans over millions of years.

The oceans themselves during this period were chemically very different from modern oceans, not just in their overall composition but in their acidity, their dissolved metal content, and their oxygen content. The early oceans probably had a pH closer to neutral or perhaps slightly acidic, rather than the slightly alkaline pH of modern seawater, because they were being continuously fed by acidic rain containing dissolved carbon dioxide and sulfur compounds from the atmosphere. The lack of free oxygen in the water oxygen wouldn't become abundant in the atmosphere and oceans for billions of years meant that metals like iron and manganese that are normally insoluble in oxygenated water could remain dissolved in early seawater in much higher concentrations than we see today.

These dissolved metals gave the oceans a different color and opacity than modern oceans, possibly appearing greenish or brownish depending on concentrations and on what other materials were suspended in the water. The oceans were also probably warmer than today, not because of ocean currents or solar heating patterns but because the entire planet was warmer, with higher atmospheric greenhouse gas concentrations and more heat escaping from the interior through volcanic activity and through the still-elevated geothermal gradient. Ocean temperatures might have ranged from perhaps 50 or 60 degrees Celsius in the warmest regions down to 30 or 40 degrees in the coolest, far warmer than modern ocean temperatures that range from roughly -2 degrees Celsius in polar regions to about 30 degrees in tropical surface waters.

The ocean depths were probably not as cold as modern deep oceans because there was less temperature stratification the water wasn't divided into distinct warm surface layers and cold deep layers separated by a sharp thermocline. Instead, the oceans were probably relatively well-mixed vertically, kept stirred by the vigorous convection driven by volcanic heat entering from below through hydrothermal vents at spreading centers and subduction zones. This mixing meant that nutrients and dissolved materials weren't trapped in the deep ocean away from the sunlit surface layers where photosynthetic organisms would eventually need them, but instead circulated throughout the water column, creating more uniform conditions from surface to depth.

The tides in these early oceans were far more dramatic than modern tides because the Moon was much closer to Earth, perhaps only 15,000 or 20,000 miles away compared to its current distance of about 240,000 miles. The gravitational pull creating tides decreases with the cube of the distance, so at closer distances the tidal forces were enormously stronger perhaps a hundred times stronger than modern tides capable of raising water levels by hundreds of feet twice per day as the Moon's gravity pulled on the oceans. These extreme tides would have created powerful currents as water rushed into and out of bays and narrow passages between landmasses, eroding coastlines rapidly and mixing ocean water thoroughly with each tidal cycle.

The day was also much shorter during this period because Earth was spinning faster, having not yet been slowed by billions of years of tidal friction with the Moon. A day might have been only five or six hours long shortly after the Moon formed, gradually lengthening over time as the Moon's gravity exerted a braking force on Earth's rotation while simultaneously pushing the Moon farther away. This rapid rotation meant that the Coriolis effect the apparent deflection of moving objects caused by Earth's rotation was stronger than today, which would have influenced atmospheric circulation patterns and ocean currents in ways that are difficult to model precisely but that certainly made weather and ocean dynamics different from what we experience on modern Earth.

The sky during this cooling period would have been gradually transforming in appearance as the atmosphere evolved. The thick, opaque, steam-filled atmosphere of the magma ocean era had given way to something clearer but still far from the transparent blue sky of modern Earth. The atmosphere still contained substantial amounts of carbon dioxide, creating a greenhouse effect that kept temperatures warm despite the dimmer Sun. Volcanic emissions continued to inject sulfur compounds and other gases that created haze layers and colored the sky in shades that probably ranged from orange to brown to gray depending on the concentration of particles and gases.

11

The Earliest Continents

Earth by around 3.8 to 3.5 billion years ago had become a recognizable water world, covered predominantly by oceans with only scattered fragments of land rising above the waves, but it was not yet the Earth we would recognize today with its familiar continents and ocean basins. The distinction between oceanic crust and continental crust was just beginning to emerge, driven by processes that would eventually create the stable landmasses we call continents, but these earliest continental nuclei were small, unstable, and fundamentally different in character from modern continents. They were being born through violence and heat, forged in the interaction between water descending into Earth's interior and molten rock rising toward the surface, assembled through collisions and volcanic activity into masses that would grow over billions of years to become the continents we know.

The process of creating continental crust is more complex than creating oceanic crust, requiring multiple stages of melting, crystallization, and chemical differentiation to transform the basaltic composition of typical mantle-derived magma into the more silicon-rich, aluminum-rich composition characteristic of continental rocks like granite and granodiorite. Continental crust is fundamentally a product of plate tectonics or proto-plate-tectonic processes, formed primarily in settings where oceanic crust is being subducted forced down into the mantle—beneath other crust, creating conditions that allow for the specialized melting processes needed to generate continental-type magmas.

When oceanic crust descends into the mantle through subduction, it carries with it any water that had infiltrated into the rocks through hydrothermal circulation at the seafloor or through reactions with seawater. This water is initially bound in minerals compounds like serpentine, chlorite, and various amphiboles that can hold significant amounts of water in their crystal structures, perhaps ten or even twenty percent water by weight. As the subducting slab of oceanic crust descends to greater depths, it encounters increasing temperatures and pressures that eventually become too extreme for these water-bearing minerals to remain stable. The minerals break down, releasing their water into the surrounding mantle wedge the region of mantle rock lying above the subducting slab.

This released water has a dramatic effect on the mantle rock it infiltrates. Pure mantle rock, without water, has a melting temperature of around 1,200 to 1,300 degrees Celsius at the pressures found in the upper mantle, and at the typical temperatures in the mantle wedge above a subducting slab, the rock would normally remain solid. But when water is added, even in small amounts of just a few percent by weight, the melting temperature drops by several hundred degrees, falling below the actual temperature of the mantle wedge and causing the rock to begin melting. This water-induced melting produces magma that is already somewhat different in composition from typical basalt, richer in silicon and lighter in density, and this magma then rises upward through the overlying crust because it's less dense than the surrounding solid rock.

As the magma rises, it may stop and accumulate in chambers beneath the surface, cooling slowly and crystallizing into coarse-grained rocks like gabbro if the composition is still relatively basaltic, or into diorite or tonalite if the composition has evolved toward more continental types. During this crystallization, different minerals form at different temperatures in a specific sequence, and the early-forming minerals typically dark minerals rich in iron and magnesium can sink to the bottom of the magma chamber or otherwise separate from the remaining liquid, leaving that liquid enriched in silicon, aluminum, sodium, and potassium. This process, called fractional crystallization, progressively changes the composition of the remaining magma toward more continental-like compositions with each stage of crystallization and separation.

The magma may also interact with the rocks it passes through or sits within, partially melting them or dissolving components from them, further modifying its composition through a process called assimilation. If the magma rises all the way to the surface and erupts, it creates volcanoes built of andesite or dacite or rhyolite volcanic rock types with increasingly high silicon content and increasingly continental-like chemistry. Over many cycles of subduction, melting, crystallization, and eruption, these processes gradually build up volumes of continental-type crust, accumulating it in the regions above subduction zones, typically forming chains of volcanic islands initially, and eventually building larger landmasses as these island chains collide and merge.

The earliest continents, or more accurately the earliest continental nuclei that would eventually grow into continents, probably formed through these processes beginning around 4.0 to 3.8 billion years ago, though pinpointing the exact timing is difficult because so little of that ancient crust has survived to the present. What we can say with confidence is that by 3.8 billion years ago, and possibly earlier, there were at least small areas of continental-type crust existing on Earth's surface, preserving chemical signatures and rock types that indicate they formed through the multi-stage processes characteristic of continental formation rather than through the simple melting and crystallization that creates oceanic crust.

The oldest surviving fragments of continental crust that we can still find on Earth today exist as small regions within several modern continents, ancient cores called cratons that have remained relatively stable and undeformed for billions of years while younger rocks accumulated around them. The most famous of these ancient cratons include portions of Western Australia, where rocks and minerals dated to 4.4 billion years ago have been found; regions of Greenland containing rocks about 3.8 billion years old; parts of Canada where rocks approaching 4.0 billion years in age exist; and areas of South Africa, Western Australia, and northern Canada where extensive sequences of rocks between 3.5 and 2.5 billion years old preserve detailed records of early continental processes.

These ancient cratons weren't originally continents in the modern sense they were much smaller, perhaps the size of large islands or small nations rather than entire continents, and they were separated from each other by vast expanses of oceanic crust and water. Scientists sometimes call these early continental nuclei by specific names based on the evidence and the reconstructions of ancient geography: Vaalbara, possibly the earliest proto-continent, might have existed around 3.6 to 3.1 billion years ago, formed from the collision of two smaller cratons, though its existence and configuration remain debated. Ur, another proposed early landmass, might have formed around 3.0 billion years ago. Kenorland might have existed around 2.7 billion years ago, assembling several cratons into a larger landmass.

But we're getting ahead of ourselves, jumping billions of years into the future when we should still be in the period around 3.8 to 3.5 billion years ago when these continental fragments were just beginning to form. During this early period, the landmasses were small, scattered, and constantly being created and destroyed through the vigorous tectonic activity driven by Earth's high internal heat. The planet was processing crust through subduction and volcanism at rates perhaps ten times faster than today, creating and recycling oceanic crust rapidly, occasionally building up small volumes of buoyant continental crust that resisted subduction and gradually accumulated into more persistent landmasses.

The formation of these early continents had several important consequences for Earth's evolution beyond simply creating land areas above sea level. Continental crust is thicker than oceanic crust typically 25 to 40 miles thick compared to oceanic crust's 5 to 7 miles—which means it stands higher above the underlying mantle, rising farther above sea level and creating regions with different elevation ranges and different exposure to weathering processes. Continental crust is also composition ally layered, with more silicon-rich rocks near the surface and more basaltic compositions at depth, creating a vertical stratification that affects how the crust responds to heating, deformation, and erosion.

Perhaps most importantly, continental crust is less dense than oceanic crust about 2.7 grams per cubic centimeter for typical continental rocks compared to 3.0 grams per cubic centimeter for oceanic basalt which makes it more buoyant and therefore much more difficult to subduct back into the mantle. Oceanic crust, once it cools and becomes old, is dense enough that it can sink into the mantle relatively easily when pushed at a subduction zone, which is why oceanic crust is constantly being created at mid-ocean ridges and destroyed at subduction zones, with no oceanic crust older than about 200 million years existing anywhere on Earth today. But continental crust resists subduction because it floats too well—when it's pushed toward a subduction zone, it tends to jam up the system rather than smoothly descending, leading to continent-continent collisions that build mountain ranges rather than allowing continued subduction.

This resistance to destruction is why continental crust, once formed, tends to persist for billions of years, gradually accumulating and growing over time as new continental crust is added through volcanism and as existing fragments collide and merge. The oldest continental rocks are billions of years old, having survived numerous cycles of heating, deformation, burial, and uplift without being completely recycled back into the mantle, whereas oceanic crust is young, constantly being renewed, with its oldest portions always being subducted and destroyed before they can grow very old.

The early continents, small and scattered as they were, profoundly altered the way Earth's surface interacted with its atmosphere and oceans. On a planet covered entirely by ocean, with only occasional volcanic islands poking above the waves, weathering processes would be limited primarily to underwater chemical reactions between seawater and the rocks of the seafloor. But once significant areas of land emerged above sea level, exposed to direct contact with the atmosphere and to the erosive power of rainfall, an entirely new suite of weathering processes became active, processes that would eventually reshape both the surface and the atmosphere in fundamental ways.

Rainfall on exposed continental rock initiates a complex series of chemical reactions collectively called continental weathering. Rainwater, even in the absence of pollutants, is naturally slightly acidic because carbon dioxide from the atmosphere dissolves in water to form weak carbonic acid, the same compound that makes carbonated beverages fizzy. When this acidic rainwater contacts silicate minerals in continental rocks—particularly minerals like feldspar, which are abundant in granite and related rocks—it reacts with them, breaking down their crystal structures and dissolving certain elements while leaving others behind.

The calcium, sodium, potassium, and magnesium in the feldspar and other silicate minerals tend to dissolve into the water, forming dissolved ions that are carried away by streams and rivers eventually reaching the ocean. The silicon and aluminum, which are less soluble under typical surface conditions, tend to remain behind as residues, often forming clay minerals fine-grained hydrated aluminosilicates that accumulate as soils and sediments. The iron present in many continental rocks oxidizes when exposed to oxygen and water, forming rust-colored iron oxide minerals that give many soils and sediments their reddish or yellowish-brown colors, though in the oxygen-poor early Earth, iron oxidation would have been less prominent and iron might have remained in more reduced forms.

Crucially, this continental weathering process consumes carbon dioxide from the atmosphere each molecule of carbonic acid that reacts with silicate minerals removes one CO_2 molecule from the atmosphere and locks that carbon away in solid form when the dissolved calcium or magnesium eventually combines with carbonate ions in the ocean to precipitate as limestone or other carbonate minerals. This represents a long-term sink for atmospheric carbon dioxide, a mechanism by which CO_2 can be removed from the atmosphere for millions or billions of years, stored in rocks rather than contributing to the greenhouse effect.

Without continents and continental weathering, this carbon dioxide removal mechanism would be much weaker—seafloor weathering can also consume CO_2, but it's less efficient because the exposed area is smaller and because the reactions proceed more slowly underwater than in the presence of acidic rain. The emergence of continents therefore gave Earth a powerful new climate regulation mechanism, a way to remove excess atmospheric carbon dioxide over long timescales, which would become increasingly important as the Sun gradually brightened and Earth needed to reduce its greenhouse effect to avoid overheating.

The weathering of continental rocks also delivered nutrients to the oceans that would eventually become important for life. Elements like phosphorus, iron, and various trace metals are relatively insoluble in seawater and don't occur in high concentrations in dissolved form, but they're essential for biological processes phosphorus for making DNA and cell membranes, iron for making proteins involved in respiration and photosynthesis, trace metals for various enzymatic reactions. Continental weathering, by breaking down rocks and releasing these elements into rivers that carry them to the ocean, provides a continuous supply of nutrients that can support biological productivity, particularly in coastal regions near river mouths where the nutrient concentrations are highest.

The physical weathering of continents, in addition to chemical weathering, began creating sediments in quantities and varieties that hadn't existed before.

Physical weathering the breaking apart of rocks through mechanical processes like freeze-thaw cycles, thermal expansion and contraction, and the grinding action of rocks transported by water or ice produces fragments ranging from tiny clay particles to sand grains to pebbles and boulders. These fragments, once liberated from the parent rock, can be transported by water, sorted by size and density, and deposited in layers to form sedimentary rocks.

The earliest sedimentary deposits formed from continental weathering probably accumulated in shallow coastal waters and on the seafloor near the small landmasses, creating layers of sandstone, mud stone, and conglomerate rocks formed from the consolidation of sand, mud, and gravel respectively. These sedimentary rocks would have been composition ally different from the volcanic and metamorphic rocks that dominated earlier periods, carrying chemical and mineral signatures of their continental source rocks and preserving information about the conditions under which they were transported and deposited.

Some of the oldest preserved sedimentary sequences on Earth, dating back to around 3.8 to 3.5 billion years ago, include rocks that appear to have been deposited in shallow water environments, possibly in coastal settings near early landmasses. These rocks contain rounded pebbles and sand grains that show evidence of having been transported by water, sorted by currents, and deposited in layered sequences that suggest changing water depths or flow conditions over time. The presence of these sedimentary rocks provides strong evidence that continents and continental weathering processes were already operating by this early date, creating the source material for these sediments and the erosive and transport mechanisms to move them around.

The tectonic processes that built these early continents operated under somewhat different conditions than modern plate tectonics, primarily because Earth's interior was significantly hotter during this period than it is today. A hotter interior meant a steeper temperature gradient from the core to the surface, which drove more vigorous mantle convection and resulted in higher heat flux through the crust. This high heat flux had several effects on how continents formed and evolved.

First, it meant that volcanic activity was more intense and widespread, producing more magma and building volcanic landmasses more rapidly than occurs today. The early continents might have grown through accretion of volcanic material at rates several times faster than modern continental growth, with massive volcanic eruptions building island arc systems that then collided and merged to form larger landmasses.

Second, the high heat flux meant that newly formed crust didn't cool and thicken as effectively as modern crust does, remaining hotter and weaker for longer periods. This made the crust more ductile, more prone to flowing and deforming rather than fracturing brittly, which affected how tectonic stresses were accommodated. Instead of building high mountain ranges through the collision of rigid crustal blocks, as happens in modern continent-continent collisions like the one that created the Himalayas, early continental collisions might have produced broader zones of deformation where the crust thickened and flowed rather than being thrust upward into sharp peaks.

Third, the rapid plate velocities driven by vigorous mantle convection meant that plates moved faster relative to each other than modern plates do, creating more frequent collisions and faster cycles of ocean basin opening and closing. The early Earth might have had a plate tectonic cycle time the period from when an ocean basin opens to when it closes through continental collision of perhaps tens of millions of years rather than the hundreds of millions of years characteristic of modern plate cycles. This rapid cycling would have made the configuration of continents and ocean basins highly unstable, constantly changing as small continental fragments moved around the globe, collided, split apart, and reassembled in different configurations.

The early continents were also probably less stable vertically than modern continents, experiencing more dramatic changes in elevation over time due to the dynamic processes affecting them. Continental crust can rise or sink vertically in response to changes in thickness, temperature, or density through a process called isostasy the tendency of Earth's crust to float at an elevation determined by its density and thickness, similar to how an iceberg floats with most of its mass submerged and only a fraction protruding above water.

12

Chemistry at the Edge of Life

Earth had transformed over the course of several hundred million years from a molten, sterile ball of rock into something far more complex and structured—a planet with solid crust and liquid oceans, with an atmosphere that had stabilized into a persistent envelope of gases, with continents beginning to emerge above the waves, with chemical cycles moving elements between rocks and water and air. The surface temperature had cooled to ranges where liquid water could exist comfortably, probably somewhere between 30 and 80 degrees Celsius depending on location and time, warm by modern standards but not lethally hot. The oceans covered most of the planet, perhaps 95 percent or more of the surface, containing dissolved minerals and salts washed from the rocks, chemically active and constantly mixing through currents and tides. The atmosphere still lacked free oxygen but contained nitrogen, carbon dioxide, water vapor, and traces of other gases, creating a reducing environment where certain types of chemistry could occur that would be impossible in an oxygen-rich atmosphere.

And in this world of rock and water and air, in the boundary zones where different environments met and interacted in tidal pools along rocky shorelines, in hydrothermal vents on the ocean floor, in shallow lakes on volcanic islands, perhaps in ice-covered ponds or in the tiny water-filled pores within rocks chemistry was becoming complicated in ways it had never been before.

Molecules were forming that were larger and more intricate than the simple compounds like water and carbon dioxide that dominated the environment, molecules with dozens or hundreds of atoms linked together in specific sequences and three-dimensional shapes, molecules that could do things beyond simply existing, that could catalyze reactions or store information or capture energy.

This was not life, not yet, but it was chemistry that was approaching the threshold of life, exploring the space of possibilities that exists between simple inorganic chemistry and the full complexity of living cells. It was chemistry that was learning, through random trial and error over millions of years, how to build the molecular machinery that life would eventually require how to make the long carbon-chain molecules that could serve as structural materials or energy storage, how to assemble the nucleotide that would carry genetic information, how to create the amino acids that would link together into proteins capable of catalyzing the chemical reactions that metabolism requires.

The exact pathways by which this prebiotic chemistry occurred, the specific locations where it happened, and the timeline over which simple molecules transformed into complex ones remain among the most uncertain and debated questions in all of science. We don't have direct evidence from this period no fossils exist from before life began, obviously, and the rocks that might have preserved chemical traces of prebiotic processes have been mostly destroyed by billions of years of geological recycling. What we have instead is inference and experimentation, theoretical calculations about what chemistry is possible under what conditions, laboratory experiments that attempt to recreate early Earth environments and see what molecules form spontaneously, and observations of organic chemistry in other settings like interstellar space or on meteorites that might tell us about the raw materials available.

The story begins with the simple molecules that were abundant on early Earth water, carbon dioxide, nitrogen, methane, ammonia, hydrogen sulfide, and smaller amounts of other compounds and asks how these could be transformed into more complex organic molecules, molecules containing carbon bonded to hydrogen, oxygen, nitrogen, sulfur, and phosphorus in chains and rings of various lengths and arrangements. Organic chemistry is fundamentally the chemistry of carbon, and carbon's ability to form stable bonds with itself and with many other elements creates an almost infinite variety of possible molecular structures, from simple two-carbon ethane to the enormous tangled polymers that make up proteins and DNA.

The transformation from simple inorganic molecules to complex organic ones requires energy, because the complex molecules are in a higher energy state than the simple ones they're built from thermodynamics doesn't favor their spontaneous formation without an energy input to drive the reactions uphill. On early Earth, several energy sources were available that could potentially drive this chemistry. Ultraviolet radiation from the Sun, far more intense than what reaches Earth's surface today because there was no ozone layer to absorb it, could break apart molecules and create reactive fragments that would recombine into new compounds. Lightning strikes, probably common in the turbulent early atmosphere, could generate tremendous electrical discharges that provided energy to drive reactions in the gases they passed through. Volcanic heat and the chemical energy stored in hot volcanic gases could power reactions, particularly in the vicinity of eruptions or hot springs. And perhaps most importantly, the chemical disequilibrium between certain environments places where reduced chemicals from Earth's interior met oxidized chemicals from the atmosphere or ocean could drive reactions through redox chemistry, using the energy released when electrons are transferred from one molecule to another.

The classic demonstration that organic molecules can form spontaneously under early Earth conditions came in 1952, when Stanley Miller, then a graduate student working with Harold Urey at the University of Chicago, created a simple apparatus that circulated a mixture of methane, ammonia, water vapor, and hydrogen through a flask where electrical sparks simulated lightning.

After running the experiment for a week, Miller analyzed the water that had accumulated in the apparatus and found that it contained several amino acids the building blocks of proteins along with other organic compounds. The amino acids had formed spontaneously from the simple starting gases, driven by the energy from the electrical discharges, demonstrating that at least some of the molecular building blocks of life could arise through purely chemical processes without requiring any biological intervention.

The Miller-Urey experiment, as it came to be known, was groundbreaking and influential, but subsequent research has complicated its interpretation. The gas mixture Miller used methane, ammonia, and hydrogen represented what scientists in the 1950s thought the early atmosphere might have contained, based on assumptions about atmospheric composition derived from studying other planets. But more recent understanding of Earth's early atmosphere suggests it probably contained more carbon dioxide and nitrogen and less methane and ammonia than Miller assumed, which changes what chemistry is possible and what products form. Experiments using more realistic atmospheric compositions still produce organic molecules, but often in lower yields and with different distributions of products than the original Miller-Urey experiment generated.

Moreover, the atmosphere might not have been the most important location for prebiotic chemistry. Many researchers now think that environments in direct contact with rocks and minerals particularly hydrothermal systems where hot water circulates through rocks, either on land as hot springs and geysers or underwater as hydrothermal vents might have been more favorable for organic synthesis than open water or the atmosphere. Mineral surfaces can act as catalysts, speeding up reactions that would otherwise occur too slowly to be significant, and they can also concentrate organic molecules from dilute solutions, bringing reactants together in close proximity where they're more likely to interact.

Hydrothermal vents on the ocean floor, where hot mineral-rich water emerges from cracks in the oceanic crust, create particularly interesting chemical environments for potentially generating organic molecules and perhaps even for the origin of life itself. These vents form where seawater percolates down through fractures in the seafloor, gets heated by the hot rocks near magma chambers beneath mid-ocean ridges or near volcanic regions, and then rises back up through the crust, emerging at the seafloor at temperatures that can reach 300 to 400 degrees Celsius in some cases, though cooler vents also exist.

The water emerging from these hydrothermal vents is chemically very different from normal seawater because it has been altered by reactions with hot rocks deep underground, picking up dissolved minerals, metals, and gases during its journey through the crust. The vent fluids are typically rich in hydrogen, methane, hydrogen sulfide, and various metal ions like iron, manganese, and zinc, creating a strongly reducing chemical environment—the opposite of the oxidizing conditions that exist in oxygenated seawater. When this hot, reduced vent fluid mixes with the cold, relatively oxidized seawater surrounding it, the chemical contrast drives reactions as electrons are transferred between molecules, releasing energy that can potentially be harnessed to build complex organic molecules or to power primitive metabolic processes.

Some modern hydrothermal vents, particularly a type called alkaline hydrothermal vents or Lost City-type vents after the Lost City vent field discovered in the Atlantic Ocean, produce fluids that are warm rather than scalding hot perhaps 40 to 90 degrees Celsius—and are alkaline rather than acidic, with high pH values around 9 to 11. These vents form when seawater reacts with certain types of mantle rocks containing minerals like olivine, producing a chemical reaction called serpentinization that generates hydrogen gas, methane, and creates alkaline fluids. The vents build tall chimneys of carbonate and silica minerals that can be hundreds of feet high, riddled with tiny pores and channels where the alkaline vent fluid mixes with seawater.

These alkaline vents are particularly interesting as potential sites for the origin of life because they create natural chemical gradients across thin mineral barriers the walls of the pore spaces in the chimney structures. On one side of the barrier is alkaline, hydrogen-rich vent fluid; on the other side is more acidic, relatively oxidized seawater. This pH gradient across a membrane, with alkaline conditions on one side and more neutral or acidic conditions on the other, resembles the proton gradients that modern cells use to generate energy through a process called chemiosmosis, where the flow of protons across membranes drives the synthesis of ATP, the universal energy currency of life.

The alkaline vent hypothesis, developed primarily by Michael Russell and his colleagues starting in the 1980s and 1990s, proposes that life might have originated in these pore spaces within alkaline vent chimneys, where natural chemical gradients could have driven primitive metabolism before cells had evolved the complex protein machinery needed to create and maintain those gradients themselves. The mineral walls of the pores could have served as the first membranes, separating different chemical environments and allowing gradients to persist, before organic membranes made of lipids evolved to take over that function. The metal sulfide minerals in the chimney walls could have acted as catalysts for reactions, playing roles similar to the metal-containing enzymes that modern cells use for metabolism.

Whether or not alkaline vents were actually the site where life originated and we may never know for certain, given the lack of direct evidence from four billion years ago they illustrate the kind of environment where the boundary between geochemistry and biochemistry might have become blurred, where purely chemical processes driven by geological energy sources might have begun producing increasingly complex molecules through pathways that would eventually be co-opted and refined by biological evolution.

But hydrothermal vents were not the only possible location for prebiotic chemistry. Shallow pools and ponds on land, particularly in volcanic regions, offered different conditions that might have been favorable for organic synthesis. Charles Darwin, in a famous letter written in 1871, speculated about life originating in "some warm little pond, with all sorts of ammonia and phosphoric salts, light, heat, electricity, etc. present," and this warm pond scenario has been elaborated and refined by subsequent researchers into several versions of what's sometimes called the primordial soup hypothesis.

The advantage of pools and ponds, compared to the open ocean, is concentration. The ocean is vast and dilute molecules dissolved in seawater are spread through enormous volumes of water, making it unlikely they'll encounter each other frequently enough to react and form larger structures. But in a pool or pond, particularly one that undergoes cycles of filling with water and then partially evaporating or drying out, molecules can become concentrated as the water volume decreases, bringing them into close contact and greatly increasing the probability of reactions occurring. Furthermore, pools exposed to the atmosphere and sunlight can access energy sources like UV radiation and atmospheric electrical discharges more directly than the deep ocean floor can.

Cycles of wetting and drying, which would naturally occur in pools near shorelines or in volcanic regions where water levels fluctuate, are particularly interesting for synthesizing certain types of molecules. Some reactions important for building biological molecules, particularly the formation of bonds between amino acids to make peptides or between nucleotide to make RNA and DNA, release water as a byproduct they're called condensation reactions or dehydration synthesis because they join molecules together by removing water. These reactions are thermodynamic ally unfavorable in liquid water, meaning they don't occur spontaneously in dilute aqueous solutions because the equilibrium favors breaking the bonds rather than making them.

But during drying phases, when water is being removed from the system through evaporation, the same reactions can proceed much more favorably, building up polymers from small molecule precursors. Laboratory experiments have shown that cycles of wetting and drying can drive the synthesis of peptides from amino acids and of RNA-like molecules from nucleotide, suggesting that environments experiencing these cycles might have been particularly important for creating the long-chain molecules that would eventually become proteins and nucleic acids.

Freezing, counter intuitively, might also have helped prebiotic chemistry. When dilute solutions freeze, the ice that forms is relatively pure water, while dissolved substances become concentrated in pockets of liquid that remain between the ice crystals a process called freeze concentration. This can increase concentrations by factors of a hundred or more, bringing molecules together and potentially facilitating reactions. Additionally, the surfaces of ice crystals can organize molecules in specific orientations, potentially catalyzing reactions or favoring the formation of certain molecular structures over others. Some researchers have proposed that life might have originated in ice-covered lakes or ponds, perhaps in polar regions or during cold periods in Earth's history, where cycles of freezing and thawing could have driven organic synthesis.

Regardless of location, the prebiotic chemistry had to solve several fundamental problems to bridge the gap between simple molecules and living cells. One problem was creating information-carrying molecules some way to store instructions for building complex structures and to transmit those instructions from one generation to the next, which is what DNA does in modern organisms. Another was creating catalysts molecules that could speed up specific chemical reactions without being consumed, allowing metabolic processes to occur at useful rates, which is what protein enzymes do. A third was creating compartments boundaries that could separate the inside of a proto-cell from the outside environment, creating a distinct chemical environment and preventing useful molecules from diffusing away, which is what lipid membranes do.

In modern cells, these three functions are handled by different types of molecules: DNA stores information, proteins catalyze reactions, and lipids form membranes. But this division of labor creates a chicken-and-egg problem for the origin of life—you need DNA to encode the instructions for making proteins, but you need proteins to copy DNA and to catalyze the reactions needed to build more DNA and proteins and membranes. How could such an interdependent system have originated when all the parts seem to require each other to function?

The leading solution to this puzzle is the RNA world hypothesis, which proposes that early life used RNA ribonucleic acid to perform both the information storage role that DNA handles today and the catalytic role that proteins handle. RNA is chemically similar to DNA, consisting of nucleotides linked together in long chains, and it can store genetic information in the sequence of its nucleotides just as DNA does. But unlike DNA, some RNA molecules can also fold into three-dimensional shapes that allow them to catalyze chemical reactions, acting as enzymes despite being made of nucleic acid rather than protein.

13

The First Signs of Life

Somewhere on early Earth, at some moment we cannot precisely identify, in some location we cannot definitively name perhaps in a hydrothermal vent system on the ocean floor, or in a tidal pool along a rocky shoreline, or in a film of water within the pores of volcanic rock chemistry crossed an invisible threshold and became biology. A collection of molecules organized itself into a system that could maintain its own structure against the tendency toward disorder, that could capture energy from its environment and use that energy to build more of itself, that could reproduce, creating copies that inherited the chemical information needed to repeat the process. This system was not alive in any sense we would recognize by looking at it it had no cells with membranes and nuclei, no DNA, no proteins as we know them, possibly not even anything we could call a distinct individual organism. But it had the essential properties that separate life from non-life: metabolism, reproduction, and the capacity for evolutionary change through variation and selection.

We have no fossils from this moment, no direct evidence of what the first living things looked like or how they functioned, because whatever structures they formed were too small, too soft, or too simple to leave traces that could survive billions of years of geological transformation. The oldest fossils we can definitively identify as biological structures that bear the unmistakable chemical or physical signatures of life date to perhaps 3.5 billion years ago, possibly 3.7 or 3.8 billion years ago if certain controversial structures are indeed biological, which means there's a gap of perhaps several hundred million years between the formation of stable oceans and continents around 4.0 billion years ago and the first preserved evidence of life. During that gap, in that dark period of Earth's history from which almost no rocks survive intact, life originated and evolved to the point where it could leave traces we can still detect.

What can we infer about this earliest life from the traces it left and from our understanding of what's necessary for life to function? First, it must have been microscopic, probably smaller than the smallest bacteria we know today, because complex multicellular organisms require sophisticated machinery for coordinating the activities of many cells, machinery that clearly took billions of years to evolve. The first living things were almost certainly single cells or even simpler than cells perhaps just self-replicating molecules enclosed in primitive membranes, entities barely qualifying as organisms at all.

Second, it must have been chemically simple compared to modern cells, which contain thousands of different types of molecules working together in extraordinarily complex networks. Modern bacteria, despite being among the simplest living organisms, possess roughly 2,000 to 5,000 different genes encoding proteins, along with sophisticated systems for reading those genes, manufacturing proteins, regulating cellular processes, sensing the environment, moving, and responding to threats. The first living things could not possibly have had this level of complexity they must have functioned with far fewer components, perhaps just a handful of different molecules capable of the most basic functions of capturing energy, building simple structures, and copying genetic information with only minimal accuracy.

Third, it must have been far less efficient than modern life at every task it performed. Modern cells are the products of billions of years of evolutionary refinement, with their biochemical pathways optimized through countless rounds of selection to operate with remarkable speed and precision. The first living things would have been slow, clumsy, and error-prone by comparison replicating their genetic information with many mistakes, metabolizing energy sources inefficiently, barely able to maintain themselves against the constant chemical degradation that affects all molecular structures. They existed at the ragged edge of what's chemically possible, just barely organized enough to count as alive, perpetually on the verge of dissolving back into the chemical soup from which they had emerged.

The metabolism of the earliest life forms the chemical processes by which they captured energy and built the molecules they needed must have been based on chemical reactions that could occur without the elaborate protein enzymes that modern cells use, because those enzymes themselves are products of evolution that wouldn't have existed yet. This suggests that early metabolism probably exploited naturally occurring chemical reactions that were already happening in the environment anyway, reactions driven by the chemical disequilibrium between different substances between reduced compounds from Earth's interior, like hydrogen or hydrogen sulfide, and oxidized compounds from the atmosphere or ocean, like carbon dioxide or sulfate.

Many researchers believe the first metabolic pathway might have been similar to what some modern microorganisms still use: combining hydrogen gas with carbon dioxide to make simple organic molecules like acetate, a two-carbon compound that can serve as both an energy source and a building block for larger molecules. This reaction, known as the acetyl-Co pathway or the Wood-Ljungdahl pathway in its modern enzyme-catalyzed form, releases energy that can be captured and used to power other reactions, and it produces organic molecules from inorganic starting materials, solving both the energy problem and the carbon problem that life faces.

The advantage of this particular metabolic pathway is that it can be catalyzed by relatively simple mineral surfaces containing iron and sulfur, particularly iron-sulfur clusters that can transfer electrons between molecules. These iron-sulfur minerals form naturally in hydrothermal systems where iron-rich rocks react with sulfur-bearing fluids, and they have catalytic properties similar to the iron-sulfur clusters found in the active sites of many ancient enzymes in modern cells. This similarity suggests a possible evolutionary pathway from geochemical catalysis by mineral surfaces to biochemical catalysis by protein enzymes containing the same metal clusters, with early organisms essentially borrowing catalytic capability from their mineral environment before evolving the ability to manufacture their own catalysts.

The earliest genetic material was almost certainly RNA or something chemically similar to it, as we discussed in the previous chapter, because RNA can both store information and catalyze reactions, making it possible for a simple organism to function with just one type of information-carrying molecule instead of requiring both DNA and proteins. But RNA has significant disadvantages compared to DNA it's less chemically stable, more prone to breaking apart through reactions with water, and more difficult to repair when damaged. An early RNA-based organism would have had to reproduce quickly and prolifically because its genetic material would have been constantly degrading, with each individual lasting perhaps only hours or days before its RNA broke down beyond repair.

This instability and the resulting high error rate in reproduction would actually have been beneficial in some ways during life's early evolution, because errors in copying genetic material are the source of variation on which natural selection acts. Modern organisms have evolved elaborate proofreading mechanisms to minimize copying errors because they've accumulated complex genetic information that needs to be preserved accurately, but early organisms with simple genomes and minimal complexity might have benefited from high mutation rates that allowed them to explore different chemical strategies rapidly, quickly finding innovations that improved their survival and reproduction.

The transition from non-living chemical systems to living organisms probably wasn't a sharp boundary but rather a gradual progression through intermediate stages that were increasingly lifelike without yet possessing all the characteristics we associate with true life. There might have been a period lasting millions of years when self-replicating molecules existed but weren't yet enclosed in membranes, making them more like molecular parasites using the chemical resources of their environment without being distinct individuals. There might have been systems with crude metabolisms and membranes but without heritable genetic information, making them alive in some sense but incapable of evolving. There might have been populations of proto-organisms with different levels of integration and complexity coexisting, with true cellular life eventually out competing or absorbing the simpler systems.

The earliest evidence we have that might represent traces of this primordial life comes from rocks in Greenland dated to approximately 3.7 to 3.8 billion years old, though the interpretation of these rocks remains controversial and debated among scientists. The rocks contain small layered structures and certain patterns in carbon isotopes the ratio of carbon-12 to carbon-13 that could potentially indicate biological activity, because living organisms preferentially use the lighter carbon-12 isotope in their metabolism, leaving characteristic isotopic signatures in organic matter and in the environment. But these same patterns could also be produced by non-biological chemical processes, making it difficult to prove definitively that they represent life rather than complex but non-living chemistry.

More convincing evidence comes from rocks in Western Australia, from a formation called the Apex chert dated to about 3.5 billion years ago, which contains microscopic structures that look remarkably like fossilized bacteria tiny filaments and spheres with sizes and shapes consistent with microbial cells. When these structures were first described in the 1990s, they were hailed as definitive proof that life existed 3.5 billion years ago, but subsequent research has raised questions about whether all the structures are genuinely biological or whether some might be mineral formations that happen to resemble cells. The debate illustrates how difficult it is to identify ancient life with certainty, especially when dealing with structures at the edge of what microscopes can resolve and when the rocks have been heated and deformed by billions of years of geological activity.

Despite these uncertainties about the oldest possible fossils, there is widespread agreement among scientists that life certainly existed by 3.5 billion years ago and probably considerably earlier, perhaps as early as 3.8 or 4.0 billion years ago, based on multiple lines of evidence that become increasingly robust for younger rocks. By 3.5 billion years ago, we find not just possible micro fossils but also stromatolites layered rocky structures built by communities of microorganisms, particularly photosynthetic bacteria that trap and bind sediment particles in mat-like layers that accumulate over time. Modern stromatolites still form today in a few locations around the world, particularly in Shark Bay in Western Australia, where we can observe living bacterial communities building these structures through their growth and mineral precipitation, providing us with direct analogs for understanding the ancient stromatolites.

The existence of stromatolites 3.5 billion years ago tells us several important things about early life. First, it indicates that life had already evolved beyond the simplest possible forms to reach a level of complexity that allowed cells to form communities, to coordinate their activities at least minimally, and to create structures visible to the naked eye even though the individual organisms were microscopic. Second, it strongly suggests that photosynthesis had already evolved, because Stromboli-building communities today are dominated by photosynthetic bacteria, and the layered structure of stromatolites is partly a consequence of photosynthetic organisms growing toward the light, creating successive layers as the community builds upward.

However, the photosynthesis that existed 3.5 billion years ago was almost certainly not the oxygen-producing photosynthesis that dominates modern Earth. Modern oxygenic photosynthesis, performed by plants, algae, and cyanobacteria, uses water as the electron source, splitting water molecules to extract electrons and releasing oxygen as a waste product. But this process is chemically difficult, requiring complex molecular machinery that took a long time to evolve. The earliest photosynthesis probably used simpler electron sources like hydrogen sulfide, hydrogen gas, or iron compounds, producing sulfur or other oxidized products instead of oxygen, in a process called anoxygenic photosynthesis that some bacteria still perform today.

Anoxygenic photosynthesis would have allowed early microorganisms to harness light energy from the Sun, giving them access to a virtually unlimited energy source compared to organisms relying solely on chemical energy from reactions between reduced and oxidized minerals. This was a revolutionary innovation that would have enabled photosynthetic organisms to proliferate wherever sunlight reached in shallow oceans, in tidal pools, perhaps even on land surfaces if they could tolerate desiccation creating the first large-scale biological communities and beginning to influence the chemistry of the oceans and atmosphere through their metabolic activities.

The evolution of photosynthesis, even the anoxygenic form, was probably not the first major metabolic innovation in life's history. Earlier organisms almost certainly existed that relied on simpler metabolisms based on reactions between chemicals in their environment metabolisms like the hydrogen-carbon dioxide reaction we discussed earlier, or metabolisms based on reactions between iron and sulfur compounds, or between methane and other carbon sources. These chemotrophic organisms, deriving their energy from chemical reactions rather than from light, probably dominated life's first hundreds of millions of years, living in environments like hydrothermal vents or in sediments where the necessary chemical gradients existed.

Different lineages of early microorganisms would have evolved different metabolic strategies, exploiting different available energy sources and different carbon sources, creating a diversifying tree of life even in these earliest periods. Some organisms evolved the ability to use different sulfur compounds, oxidizing hydrogen sulfide and using the energy to fix carbon dioxide. Others evolved to use iron, oxidizing ferrous iron to ferric iron and capturing the released energy. Still others evolved methanogenesis, producing methane as a metabolic byproduct from carbon dioxide and hydrogen. Each innovation opened up new ecological niches, new environments where organisms with the right metabolism could thrive, driving further diversification and increasing the total biomass of life on Earth.

By 3.5 billion years ago, Earth had been transformed from a lifeless world of rock and water into a living planet, though the changes were still subtle and largely invisible to any hypothetical observer. The atmosphere was still oxygen-free, the continents still small and scattered, the oceans still dominated by inorganic chemistry. But in certain locations shallow coastal waters, tidal zones, hydrothermal systems life had established itself, creating islands of biological activity in a largely sterile world. Microorganisms were processing chemicals, converting inorganic carbon into organic matter, releasing metabolic byproducts that began to alter local chemistry, building physical structures like stromatolites that changed how sediments accumulated and how light and nutrients were distributed.

Life at this stage was simple, microscopic, and invisible to any human observer who might have stood on an early shore, but it represented a fundamental transition in Earth's history the point at which the planet gained the capacity to not just exist but to evolve, to change itself through the activities of organisms that could adapt to new conditions and discover new ways of extracting energy from their environment. The chemical evolution that had built complex molecules from simple ones had given way to biological evolution, where variation, heredity, and selection could drive changes far faster than chemistry alone could achieve.

The exact timing of life's origin remains uncertain and may never be known with precision, lost in the gap between the earliest preserved rocks and the formation of the oceans. But whenever it happened whether 4.0 billion years ago or 3.9 or 3.8 billion years ago it happened relatively quickly after the conditions for life became possible, suggesting that the transition from chemistry to biology, while requiring very specific conditions and probably a long series of incremental steps, is not impossibly unlikely. Earth became habitable, and within perhaps a few hundred million years at most, it became inhabited, seeded with microscopic life that would spend the next three billion years slowly reshaping the planet into something even more extraordinary.

14

Earth Becomes a Living Planet

Life had emerged, established itself in scattered environments across the planet, and begun the slow work of diversification and adaptation that would eventually fill nearly every available niche on Earth's surface. But for billions of years after its origin, life remained microscopic, simple in structure, and limited in its influence on the planet's physical and chemical systems. The Earth at 3.5 billion years ago would have looked from space much as it did shortly after the oceans formed a blue world covered mostly by water, with small brown or gray continents poking above the waves, an atmosphere still lacking oxygen, a surface still occasionally scarred by large impacts from asteroids and comets left over from the solar system's formation. An observer with no sophisticated instruments might have concluded the planet was lifeless, seeing no obvious signs of biological activity, no forests or grasslands, no visible movement of animals, nothing green.

Yet beneath this apparently sterile surface, in the oceans and in the thin films of water coating rocks and sediments, life was not just persisting but actively beginning to reshape the world that had created it. The relationship between life and Earth was becoming reciprocal where initially Earth's geology and chemistry determined where and how life could exist, increasingly life's activities were influencing the geology and chemistry, creating feedback loops that would eventually transform the planet's surface into something that could not have existed without biology. Earth was crossing another threshold, moving from a planet that happened to harbor life to a planet whose surface conditions were substantially controlled by life, becoming what we might call a living planet rather than just a planet with life on it.

The transformation happened slowly, incrementally, through processes so gradual that millions of years would pass with little visible change, yet the cumulative effect over hundreds of millions and eventually billions of years was profound. The mechanisms were chemical, physical, and evolutionary, operating simultaneously and influencing each other in complex ways that we're still working to fully understand.

Chemically, living organisms were beginning to alter the composition of the oceans and the atmosphere through their metabolic activities, taking in certain substances and releasing others, accumulating some elements in their bodies and excreting others as waste. Every living cell is a chemical factory, constantly processing molecules capturing carbon from carbon dioxide or from organic compounds, incorporating nitrogen from ammonia or nitrate, using phosphorus to build nucleic acids and cell membranes, requiring sulfur for proteins and other molecules, taking up iron and other metals for enzymes. When cells die and decompose, some of these elements are returned to the environment in different chemical forms than they were taken up in, gradually changing the chemical inventory of the ocean and atmosphere.

The most dramatic example of life altering Earth's chemistry the rise of oxygen in the atmosphere still lay far in the future at 3.5 billion years ago, probably not occurring until around 2.4 billion years ago in what's called the Great Oxidation Event. But even before oxygen accumulated, life was influencing atmospheric and oceanic chemistry in subtler ways. Methanogenic microorganisms, producing methane as a byproduct of their metabolism, were releasing this potent greenhouse gas into the oceans and atmosphere, possibly maintaining substantial atmospheric methane concentrations that helped keep Earth warm despite the dimmer young Sun. Sulfate-reducing bacteria were converting sulfate in seawater to hydrogen sulfide, changing the oxidation state of sulfur in the oceans and influencing the precipitation of iron and other minerals.

Photosynthetic organisms, even those performing anoxygenic photosynthesis that didn't release oxygen, were pulling carbon dioxide from the ocean and atmosphere and incorporating it into organic matter, effectively removing CO_2 from the inorganic carbon cycle at least temporarily. Some of this organic carbon was buried in sediments before it could decompose, representing a long-term sink for carbon dioxide that would gradually reduce atmospheric CO_2 concentrations over millions of years. This biological carbon pump created a new pathway for removing greenhouse gases from the atmosphere beyond the purely geochemical weathering processes we discussed earlier, potentially allowing Earth to maintain habitable temperatures even as the Sun brightened over geological time.

Physically, organisms were beginning to alter the structure and behavior of Earth's surface in ways that affected geological processes. Microbial mats dense communities of microorganisms growing on sediment surfaces stabilized those sediments, making them less likely to be eroded by currents or waves, changing how sediment was transported and deposited. Stromatolites, those layered structures built by photosynthetic bacterial communities, created new types of sedimentary rocks with textures and structures that couldn't form through purely physical and chemical processes, adding biological sedimentary structures to the geological record for the first time.

Some microorganisms precipitated minerals as part of their metabolism or through their influence on local chemistry, creating biogenic minerals minerals formed by biological activity rather than purely geological processes. Iron-oxidizing bacteria, for example, could oxidize dissolved ferrous iron in the water, causing ferric iron compounds to precipitate out as rust-colored sediments. Calcium carbonate-precipitating bacteria could influence the formation of limestone, though the massive carbonate platforms that would later dominate tropical shallow seas probably didn't develop until more complex organisms evolved. Even the simple presence of organic matter cell walls, extracellular slimes, dead cells in sediments changed how those sediments compacted and cemented into rock, influencing the texture and properties of sedimentary rocks formed in environments where microbial life was present.

Evolutionary, life was diversifying into an increasingly wide range of forms adapted to different environments and different metabolic strategies, creating a rich ecosystem of microorganisms interacting with each other and with their environment in complex ways. The universal common ancestor of all life on Earth the organism or population of organisms from which all modern life descended had given rise to multiple lineages that would eventually differentiate into the three great domains of life we recognize today: Bacteria, Archaea, and Eukarya, though the eukaryotes (organisms with nuclei and complex internal structures, including all plants, animals, and fungi) wouldn't appear until much later, perhaps two billion years ago.

The early diversification of bacteria and archaea, occurring during this period from roughly 3.5 to 3.0 billion years ago, explored many of the fundamental metabolic possibilities available to microscopic life. Different lineages evolved to extract energy from different chemical reactions, to inhabit different temperature ranges from near-freezing waters to boiling hot springs, to tolerate different pH ranges from extremely acidic to extremely alkaline, to survive with or without oxygen once oxygen began to appear. This metabolic and environmental diversity meant that life was gradually filling more and more of the available ecological space on Earth, moving from being present in just a few favorable environments to being nearly ubiquitous anywhere liquid water existed.

The evolution of more efficient metabolisms allowed organisms to extract more energy from available resources, supporting larger populations and more complex communities. The evolution of more sophisticated genetic systems including the transition from RNA to DNA as the primary genetic material, and the development of more accurate replication and repair mechanisms allowed organisms to maintain larger genomes encoding more genes, enabling greater biochemical sophistication. The evolution of cooperative behaviors and symbiotic relationships, where different organisms lived in close association and mutually benefited, created new possibilities for metabolic integration and specialization that wouldn't have been possible for isolated individual cells.

One of the most consequential evolutionary innovations during this period was almost certainly the evolution of oxygenic photosynthesis, though pinpointing when this occurred remains challenging and controversial. Oxygenic photosynthesis using water as the electron source for photosynthesis and releasing oxygen as a waste product is performed today by cyanobacteria and by the chloroplasts in plants and algae, which are themselves descended from ancient cyanobacteria that were engulfed by other cells.

The evolution of oxygenic photosynthesis was not a simple, single-step innovation but rather required the assembly of complex molecular machinery capable of performing one of the most chemically demanding reactions in all of biology splitting water molecules to extract electrons. Water is an extraordinarily stable molecule, which is why it's so abundant and persistent on Earth's surface, but that very stability makes it difficult to break apart. Splitting water requires more energy than any other electron source used in anoxygenic photosynthesis, demanding that the photosynthetic apparatus capture light energy at two different wavelengths simultaneously and use both to drive the reaction, a feat accomplished by linking two separate light-capturing systems together in what's called the Z-scheme of photosynthesis.

The evolutionary path to oxygenic photosynthesis probably involved the gradual modification and combination of simpler anoxygenic photosynthetic systems that already existed in different bacterial lineages, brought together through horizontal gene transfer the movement of genetic material between organisms that aren't parent and offspring or through the fusion of two different types of photosynthetic organisms, or through some combination of genetic innovation and borrowing. The details remain obscure because the events occurred billions of years ago and left no direct fossil record, only the molecular signatures preserved in the genes of modern organisms that we can analyze and compare to infer evolutionary relationships and timelines.

Whenever oxygenic photosynthesis evolved whether as early as 3.5 billion years ago as some controversial evidence suggests, or somewhat later around 3.0 to 2.8 billion years ago as other interpretations indicate it represented a fundamental expansion of life's power and reach. For the first time, organisms could use the most abundant substance on Earth's surface, water, as a resource for photosynthesis, rather than being limited to more exotic and localized chemicals like hydrogen sulfide or ferrous iron. This meant photosynthetic organisms could thrive anywhere sunlight and water were both present essentially the entire surface of the sunlit ocean, shallow pools and lakes, wet surfaces on land making photosynthesis available across most of Earth's surface rather than being restricted to particular environments with special chemistry.

The oxygen released by oxygenic photosynthesis didn't immediately accumulate in the atmosphere, even though the process may have begun billions of years before the Great Oxidation Event at 2.4 billion years ago. Initially, all the oxygen produced by cyanobacteria would have been consumed by reactions with reduced substances in the environment dissolved iron and other metals in the oceans, reduced gases like methane and hydrogen in the atmosphere, organic matter from dead cells, reduced minerals in rocks. Only after these oxygen sinks had been saturated, after the oceans and atmosphere had been pushed from a reducing state toward a more neutral or oxidized state, could free oxygen finally begin accumulating in significant quantities.

This process of filling the oxygen sinks probably took hundreds of millions or even a billion years, during which oxygen concentrations remained extremely low despite ongoing production by photosynthetic organisms. The oceans during this period would have been chemically stratified, with oxygenated surface waters where photosynthesis occurred overlying deeper anoxic waters where the oxygen had been consumed before it could mix downward. This stratification created distinct chemical environments at different depths oxidized near the surface, reduced in the depths allowing different types of organisms adapted to different oxygen levels to coexist in the same ocean.

The eventual accumulation of oxygen in the atmosphere around 2.4 billion years ago an event we're jumping ahead to here but which grew directly from the processes we're describing would transform Earth more radically than any event since the formation of the oceans. Oxygen was toxic to most organisms that had evolved in anoxic conditions, lacking the biochemical machinery to handle oxygen's powerful reactivity and to protect their cellular components from oxidative damage. The rise of oxygen probably caused a mass extinction of organisms adapted to the anoxic world, wiping out lineages that couldn't adapt quickly enough to the changing conditions or retreat to environments where oxygen remained scarce.

But oxygen also opened up new evolutionary possibilities for organisms that could tolerate it or use it. Aerobic respiration using oxygen to break down organic matter and extract far more energy than can be obtained through anaerobic metabolism became possible, allowing organisms to grow faster, build larger cells, and support more complex structures and behaviors. The oxygen in the atmosphere formed an ozone layer in the upper atmosphere that shielded the surface from destructive ultraviolet radiation, making it safer for life to exist in shallow water and eventually on land without being sterilized by UV exposure. The oxidized chemistry of the surface environment changed how minerals weathered and how sediments formed, creating new types of rocks like red beds sedimentary rocks colored by oxidized iron that don't exist in the earlier rock record.

But we're moving too far forward in time. Let's return to the period we've been discussing, from roughly 3.5 to 3.0 billion years ago, when life had established itself but had not yet fundamentally altered the atmosphere's composition. During this era, sometimes called the Archean Eon by geologists, Earth's surface was an alien world by modern standards, yet it was beginning to show the signs of biological transformation that would eventually make it recognizable as a living planet.

The continents were growing, slowly accumulating through the volcanism and tectonic processes we discussed in earlier chapters, building up volumes of continental crust that stood above sea level and provided environments different from the submarine settings where oceanic crust formed. By 3.0 billion years ago, perhaps 5 to 10 percent of the modern continental volume had formed, creating scattered landmasses that might have resembled large volcanic islands or small archipelagos more than proper continents. These early continents were probably composed mainly of granite and related rocks, lighter in color and less dense than the basaltic oceanic crust, standing higher topographically and undergoing active weathering where exposed to air and rainfall.

The weathering of these continents was delivering nutrients to the oceans phosphorus, iron, trace elements that could support biological productivity, creating coastal zones and shallow shelf environments where nutrients were abundant and sunlight could penetrate to the bottom, ideal conditions for photosynthetic microbial communities. The continental shelves would have been hotspots of biological activity, with stromatolites and microbial mats covering the seafloor in shallow water, processing enormous quantities of carbon and other elements, creating locally oxygenated environments even though the atmosphere as a whole remained oxygen-free.

The ocean itself was probably quite different from modern oceans in its physical structure and circulation. The lack of polar ice caps Earth was likely ice-free or nearly ice-free during much of the Archean because of high greenhouse gas concentrations maintaining warm temperatures even at high latitudes meant there was no formation of cold, dense bottom water sinking from polar regions and driving deep ocean circulation as occurs today. Instead, ocean circulation might have been driven mainly by wind and by temperature differences between equatorial and higher latitude surface waters, creating a less vigorous overturning circulation than modern oceans display.

The ocean chemistry was certainly very different, not just in its lack of oxygen but in its concentrations of dissolved metals and its overall chemical balance. The Archean ocean was probably slightly acidic or neutral rather than alkaline like the modern ocean, with much higher concentrations of dissolved iron, manganese, and other transition metals that precipitate out in the presence of oxygen but remain dissolved in anoxic waters. The high iron content would have given the water a greenish tint in areas where iron concentrations were highest, quite different from the blue clarity of oxygenated modern seawater.

Some of the dissolved iron in the Archean ocean was being precipitated by the actions of iron-oxidizing bacteria and by the oxygen locally produced by photosynthetic organisms, creating iron-rich sediments that eventually lithified into banded iron formations distinctive sedimentary rocks with alternating layers of iron oxides and chert (silica-rich rocks) that are found in ancient rocks around the world and represent one of the clearest signatures of the transition from an anoxic to an oxic world. The banding in these formations, with iron-rich layers alternating with iron-poor layers, might reflect seasonal or longer-term cycles in ocean chemistry and biological productivity, recording oscillations in the Archean ocean's conditions that we can still read billions of years later in the rock record.

The atmosphere during this period, as we've mentioned, lacked oxygen but contained substantial amounts of methane, possibly at concentrations of hundreds or even thousands of parts per million compared to the current atmospheric level of less than two parts per million. This high methane concentration, maintained by methanogenic microorganisms producing methane faster than it could be destroyed by chemical reactions, created a greenhouse effect strong enough to keep Earth warm despite the Sun being perhaps 20 to 25 percent dimmer than today. The atmosphere might have had a hazy appearance from organic aerosols formed by photo chemical reactions breaking down methane in the upper atmosphere, creating tiny particles that would have scattered sunlight and given the sky a different color and brightness than the modern blue sky.

15

EPILOGUE

The Story Continues

The beginning of Earth is, in many ways, the beginning of everything we know. Every mountain range that punctuates a continent's skyline, every grain of sand on every beach, every drop of water in every ocean, every breath of air we draw into our lungs all of it emerged from the processes we've traced through these chapters, from the collapse of a cloud of gas and dust four and a half billion years ago to the slow, patient work of chemistry learning to become biology.

When you hold a rock in your hand, you're touching material that may have cycled through Earth's interior multiple times, melted and solidified and melted again, thrust up into mountains and worn down by rivers, transformed by heat and pressure into new mineral forms. The atoms in that rock are older than Earth itself forged in stars that died before our solar system existed, scattered through space by supernovae, drifting through the galaxy for billions of years before being gathered into the swirling disk that would become our planet.

When you stand by the ocean and watch waves break against the shore, you're witnessing water that has been part of Earth for most of its history, water that may have been locked in minerals during the planet's formation, released by volcanic eruptions during the magma ocean phase, rained from skies onto a surface too hot to hold it, evaporated and condensed countless times in an endless cycle. That water has been through hydrothermal vents kilometers beneath the sea, has been split apart by microorganisms performing photosynthesis and reassembled by others performing respiration, has carved canyons and filled lakes and frozen into glaciers and melted again through billions of years of geological and biological activity.

When you breathe, you're taking in air that is itself a product of Earth's living history. The nitrogen you breathe 78 percent of every breath was probably released from Earth's interior early in its history through volcanic out gassing, accumulated in the atmosphere because it's chemically inert and doesn't react with surface materials, and has cycled through countless organisms that incorporated it into proteins and nucleic acids before releasing it again. The oxygen that makes up 21 percent of every breath didn't exist in the atmosphere until life created it, produced as waste by photosynthetic organisms over billions of years, accumulating slowly as Earth's chemistry shifted from reducing to oxidizing, eventually reaching concentrations high enough to support complex, energetic organisms like animals.

The story we've told in these chapters is incomplete in several important ways. We've focused on the first billion years of Earth's history, from the formation of the solar system around 4.6 billion years ago to the establishment of life around 3.5 billion years ago and the beginnings of biological transformation of the planet around 3.0 billion years ago. We've left untold the subsequent three and a half billion years during which that early microbial world would evolve into something unimaginably more complex the rise of oxygen transforming the atmosphere and ocean chemistry around 2.4 billion years ago; the evolution of eukaryotic cells with nuclei and organelles around 2.0 billion years ago; the first appearance of multicellular organisms perhaps 800 million to 600 million years ago; the explosion of animal diversity in the Cambrian period 540 million years ago; the colonization of land by plants and animals between 500 and 400 million years ago; the rise and fall of dinosaurs; the emergence of mammals and eventually primates and humans.

Each of these later chapters in Earth's history deserves its own book, its own detailed exploration of the processes and events that shaped life's evolution and Earth's transformation. But they all build on the foundation established in these earliest eons the formation of a planet with the right mass and composition and orbital distance to maintain liquid water, the development of plate tectonics that recycles crust and regulates atmospheric chemistry, the establishment of a magnetic field that shields the surface from harmful solar radiation, the origin of life and its early diversification into metabolically diverse lineages.

Earth's story is also part of a larger story the story of the universe itself, of matter organizing itself into increasingly complex structures over billions of years. We are the universe's way of knowing itself, made from atoms that have been through stellar cores and supernova explosions, assembled by geological and biological processes into arrangements complex enough to contemplate our own origins. The fact that we can reconstruct Earth's beginning from fragmentary evidence preserved in ancient rocks and from theoretical understanding of planetary formation is itself remarkable, a demonstration of how far matter's self-organization has progressed from the simple atoms of the early universe to conscious beings capable of understanding their own history.

And Earth's story continues. The planet is not frozen in its current state but constantly changing, evolving, responding to both internal geological processes and external influences including, for the first time in its history, the deliberate actions of a technologically advanced species that has gained the ability to alter planetary chemistry and climate on scales comparable to geological processes. We live in a unique moment when human activity is influencing Earth's future trajectory in ways we're still learning to understand and control.

The processes that formed Earth accretion, differentiation, out gassing, cooling, weathering, the origin and evolution of life operated over timescales of millions and billions of years, far longer than human lifetimes or even human civilizations. They created a planet exquisitely balanced, with carbon cycles and water cycles and nutrient cycles that maintained habitable conditions for billions of years despite dramatic changes in solar luminosity, continental configuration, and biological innovations. Understanding these deep processes gives us perspective on our current moment, helping us recognize both the resilience of Earth's systems, which have weathered far greater catastrophes than anything humans have caused, and their potential fragility, because the processes that could restore balance operate on geological timescales while human-caused changes are occurring on timescales of decades or centuries.

The story of Earth's beginning reminds us that planets are not static objects but evolving systems, shaped by the interplay of physical, chemical, and biological processes acting over vast spans of time. It reminds us that we are not separate from nature but products of it, made from the same materials and subject to the same laws that governed Earth's formation and evolution. It reminds us that our planet's habitability is not guaranteed or permanent but depends on conditions that could change, and that we now have the knowledge and potentially the wisdom to act as stewards of a planetary system that took billions of years to create the conditions we depend on.

The beginning of Earth was violent and chaotic impacts, magma oceans, unstable atmospheres, harsh radiation, toxic chemistry. From that chaos emerged order, structure, complexity, and eventually life and consciousness. The journey from that beginning to where we stand now, able to understand and appreciate the journey, is one of the most profound stories we can tell, connecting us to the deepest past and to the largest scales of space and time.

Earth's story, which began in fire and darkness, continues in the present as a living, breathing, constantly changing world. Every sunset, every storm, every spring rain, every new life born all of it is part of the same unbroken chain of events that started with a cloud collapsing in space billions of years ago. We are not observers of this story from outside but participants within it, our own existence woven into the fabric of Earth's ongoing evolution.

The planet that began as nothing more than colliding rocks and dust has become a world of extraordinary beauty and complexity, home to millions of species living in environments ranging from the deepest ocean trenches to the highest mountain peaks, from frozen poles to steaming tropical rain forests. The chemistry that began with simple reactions between volcanic gases and rainwater has evolved into the intricate biochemistry of cells, the orchestrated interactions of ecosystems, the emergent properties of a biosphere that influences and is influenced by geological and atmospheric processes.

This is Earth's story a story of transformation, of potential becoming actual, of chaos organizing itself into order. It's a story still being written, with chapters yet to come that we can influence through our choices and actions. Understanding where we came from, how this planet formed and evolved and gave rise to us, helps us understand where we might go and what responsibilities we bear as the first species capable of consciously shaping planetary destiny.

The beginning of Earth was four and a half billion years ago, in fire and violence and darkness. But in a deeper sense, Earth's beginning continues with every sunrise, every birth, every moment of growth and change. The story never really ends it just keeps unfolding, adding new chapters to a narrative that started with stardust and may lead to destinations we can't yet imagine.

www.ingramcontent.com/pod-product-compliance
Lightning Source LLC
LaVergne TN
LVHW022051200926
848481LV00003B/868
* 9 7 9 8 2 3 1 2 0 5 3 7 0 *